This book belongs to

_____

_____

_____

O.D. KIDS

- Copyright 2020/2021 ™ © -

Copyright of this website and its contentskharaz for Printing All rights reserved.Redistribution or reproduction ofpart or all of the content in any form is prohibited

MY KINDERGARTEN MATH SKILLS ACTIVITY WORKBOOK

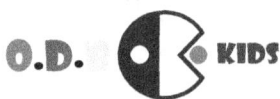
O.D. KIDS

- Copyright 2020/2021 ™ © -
Copyright of this website and its contentskharaz for Printing All rights reserved.Redistribution or reproduction ofpart or all of the content in any form is prohibited

A Guide for Parents

Welcome, parents! One of the most important gifts we can give our children is to help them learn to read and write so that they can succeed in school and beyond. Confident, active readers are able to use their reading skills to follow their passions and curiosity about the world. We all read for a purpose: to be entertained, to take a journey of the imagination, to connect with others, to figure out how to do something, and to learn about history, science, the arts, and everything else. Learning to read is complex. Children don't learn one reading-related skill and then move on to the next in a step-by-step process. Instead, they are learning to do many things at the same time: decoding, reading with comfortable fluency, absorbing new vocabulary, understanding what the text says, and discovering that reading is pleasurable and builds knowledge about the world. We hope this guide will give you a better understanding of what it takes to learn to read (and write) and how you can help your children grow as readers, writers, and learners!

INDEX

5-14 : Tracing the number

15-19 : Counting Objects

20-25 : Counting by 2's

26-30 : Identifying even / odd numbers 1-20

31-34 : Number Chart from (half-full)

35-40 : Numbers as words

41-45 : Counting Objects

46-49 : Counting practice - before / after

50-55 : Identifying tens and ones

56-61 : Combining tens and ones

62-90 : correction

Tracing the number 1 (one)
Kindergarten Numbers & Counting Worksheet

Practice tracing and printing the number 1 (one):

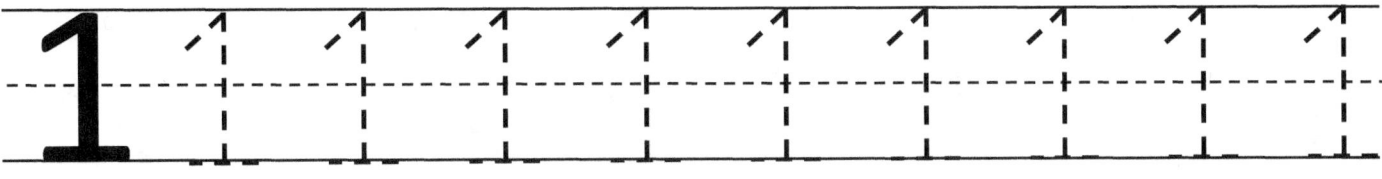

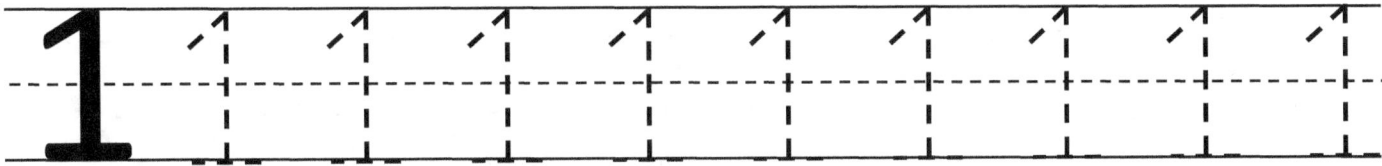

Count the map:

Circle the number 1

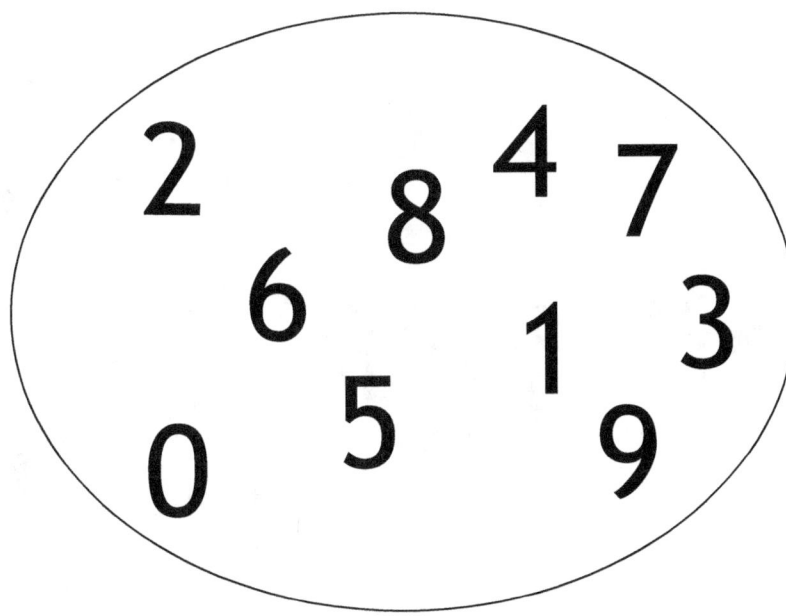

Tracing the number 2 (two)
Kindergarten Numbers & Counting Worksheet

Practice tracing and printing the number 2 (two):

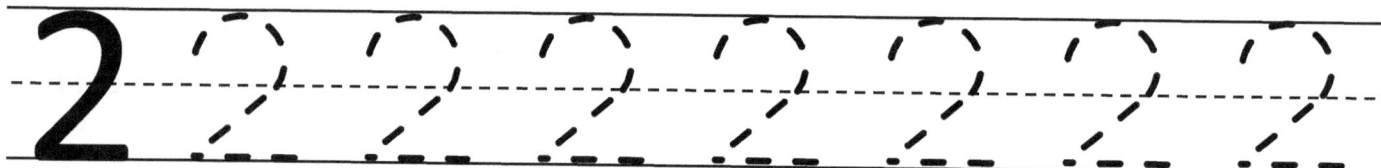

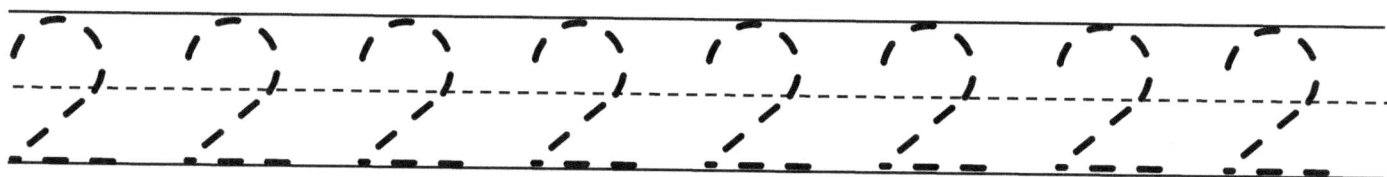

Count the Erasers: Circle the number 2

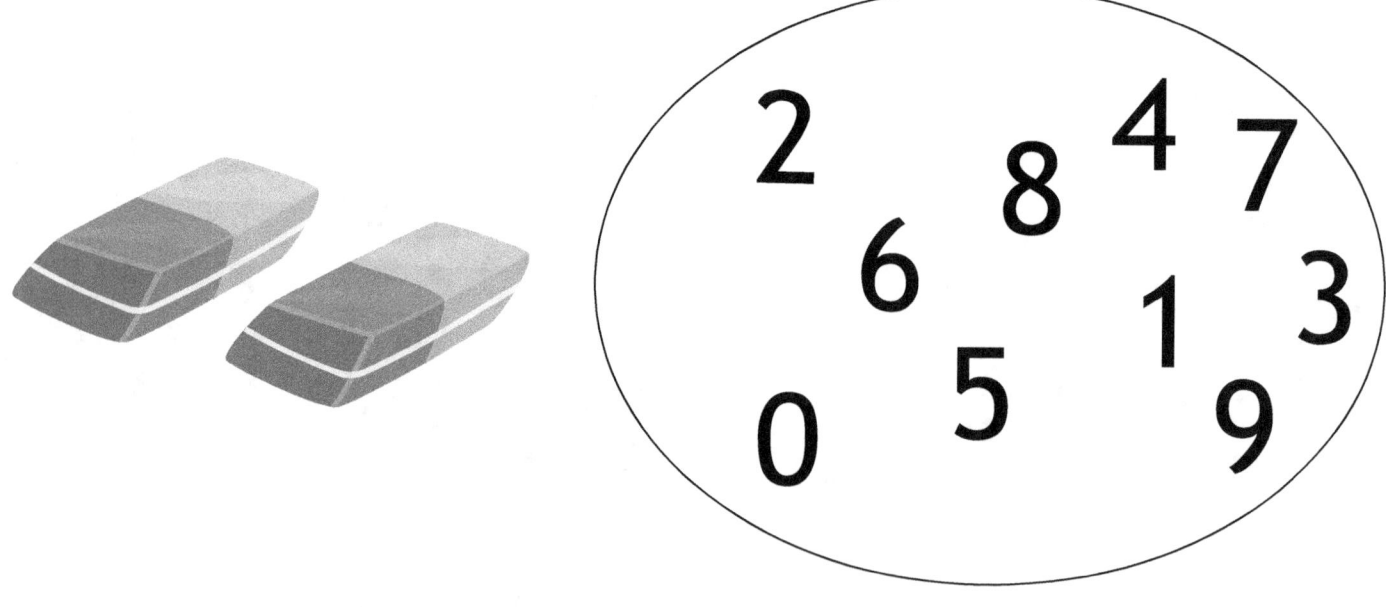

Tracing the number 3 (three)
Kindergarten Numbers & Counting Worksheet

Practice tracing and printing the number 3 (three).

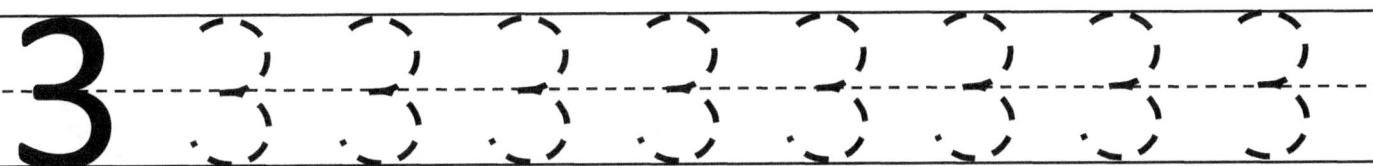

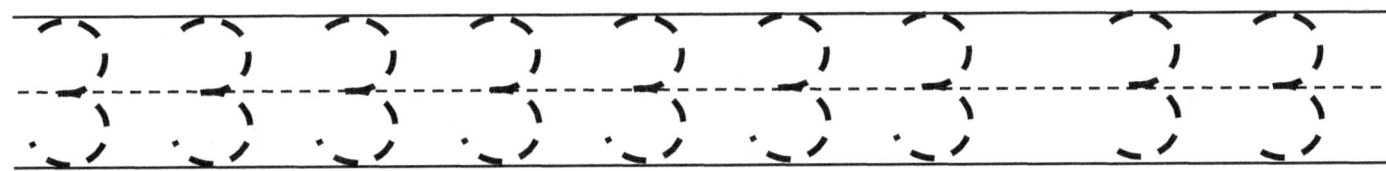

Count the balloons:

Circle the number 3

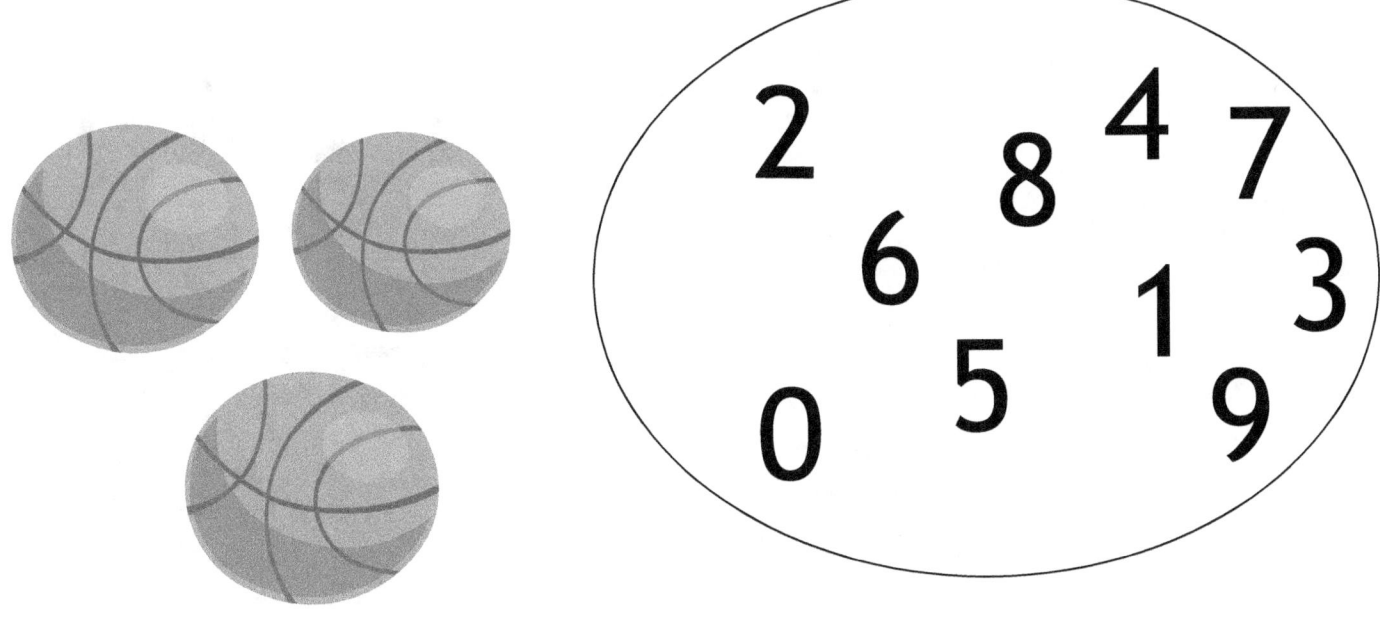

Tracing the number 4 (four)
Kindergarten Numbers & Counting Worksheet

Practice tracing and printing the number 4 (four).

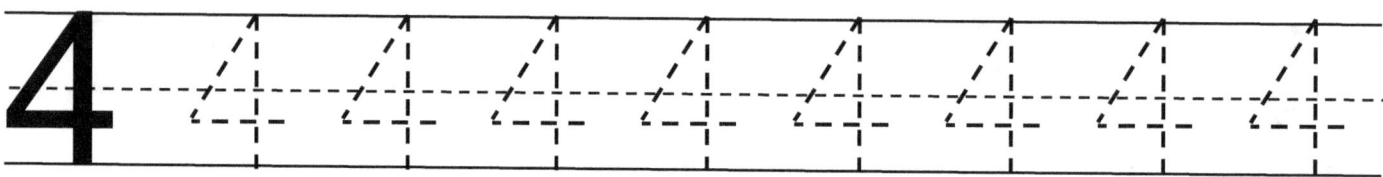

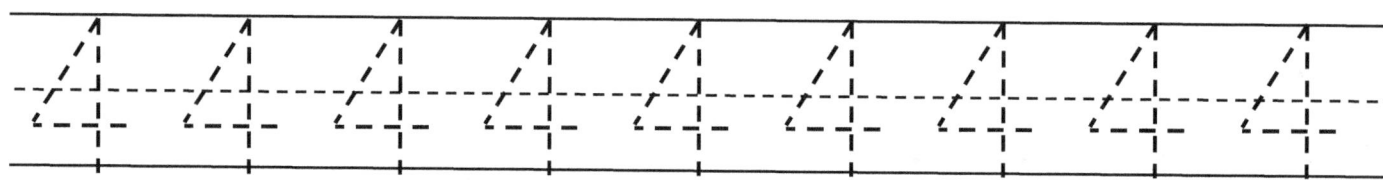

Count the ruler :

Circle the number 4:

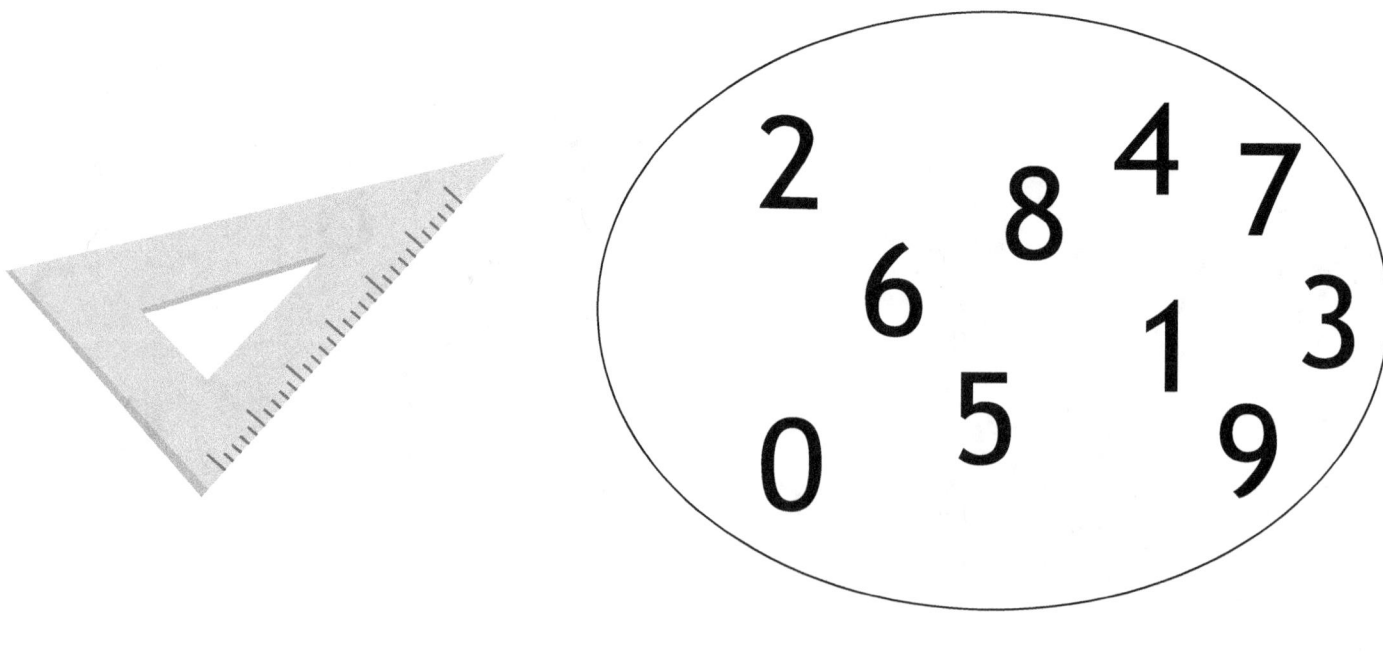

Tracing the number 5 (five)

Kindergarten Numbers & Counting Worksheet

Practice tracing and printing the number 5 (five).

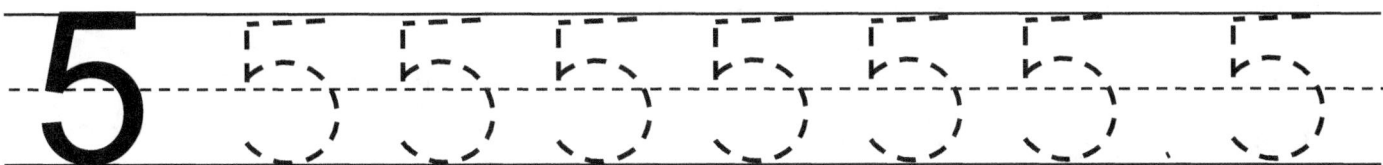

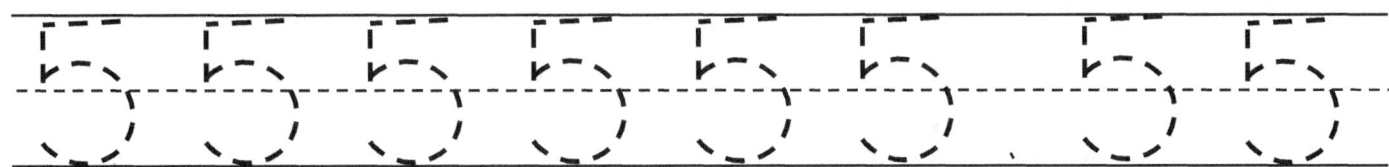

Count the School bag:

Circle the number 5:

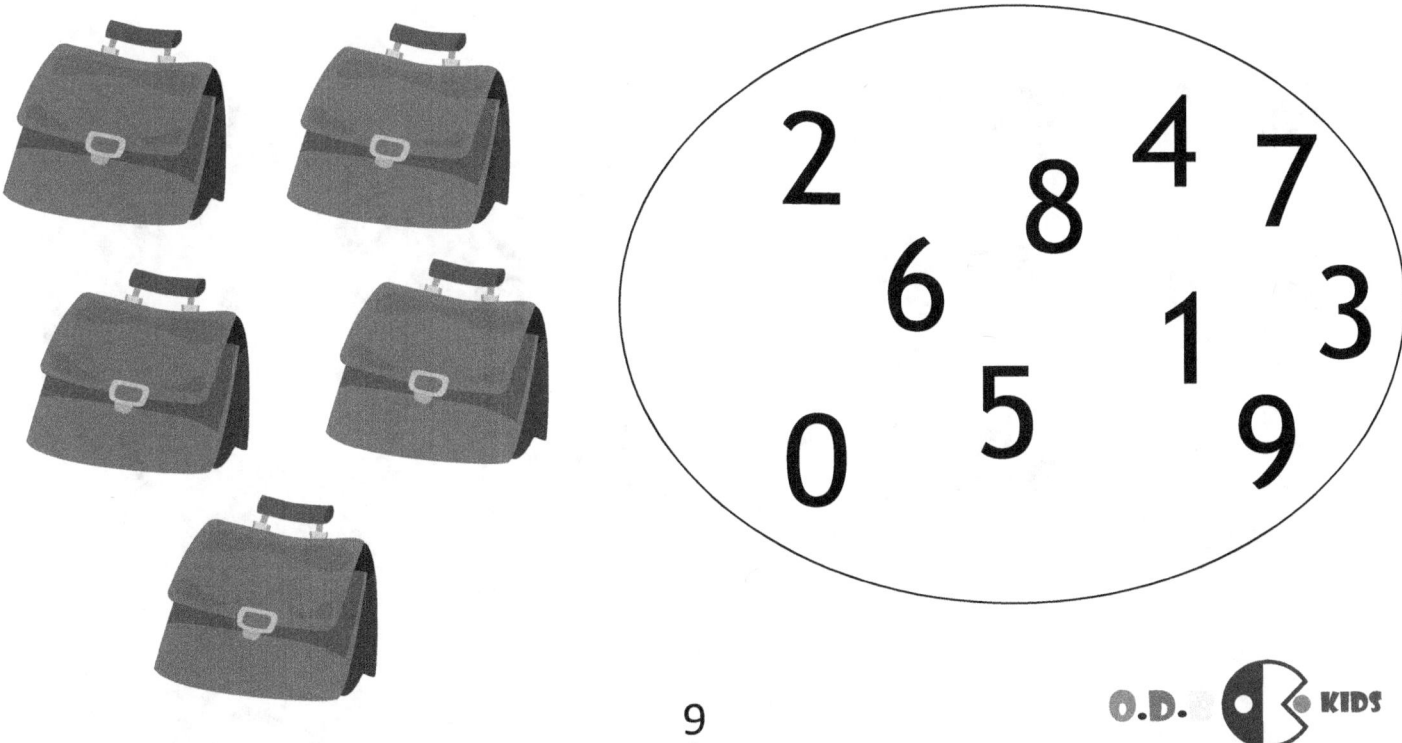

Tracing the number 6 (six)

Kindergarten Numbers & Counting Worksheet

Practice tracing and printing the number 6 (six).

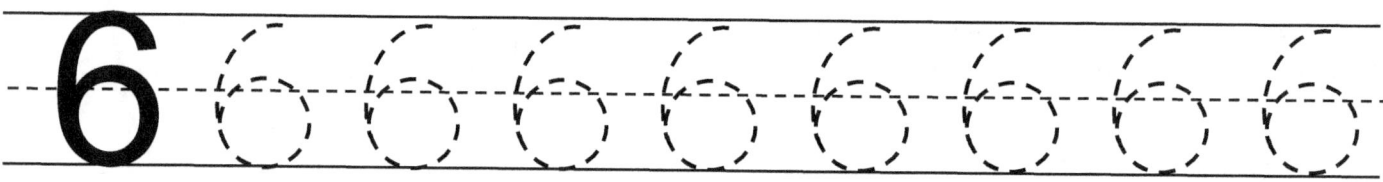

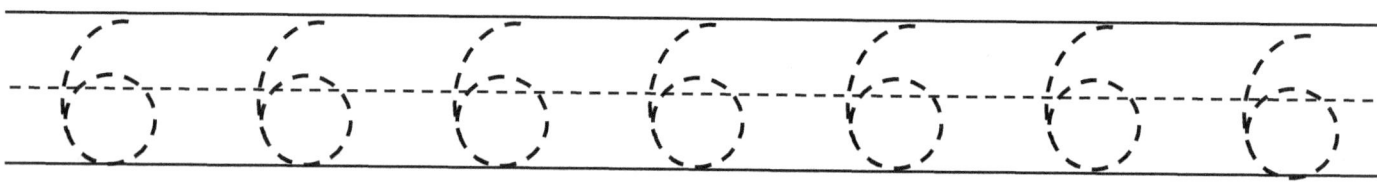

Count the pizza books :

Circle the number 6:

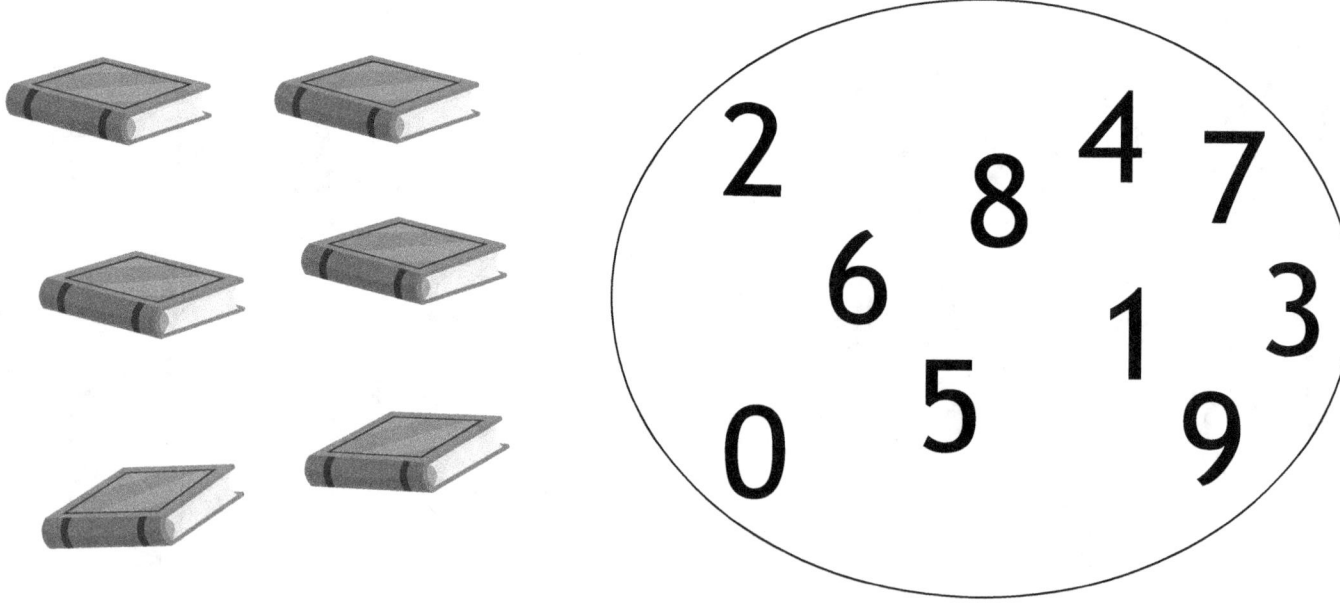

Tracing the number 7 (seven)

Kindergarten Numbers & Counting Worksheet

Practice tracing and printing the number 7 (seven).

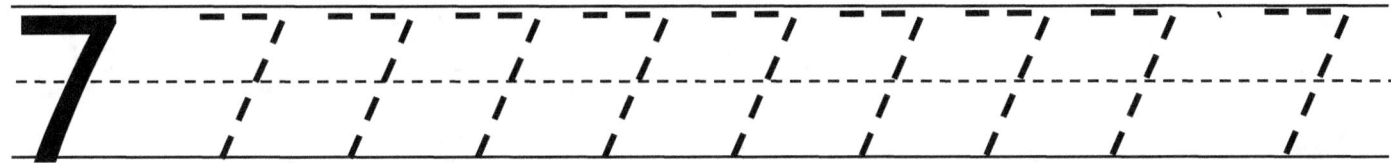

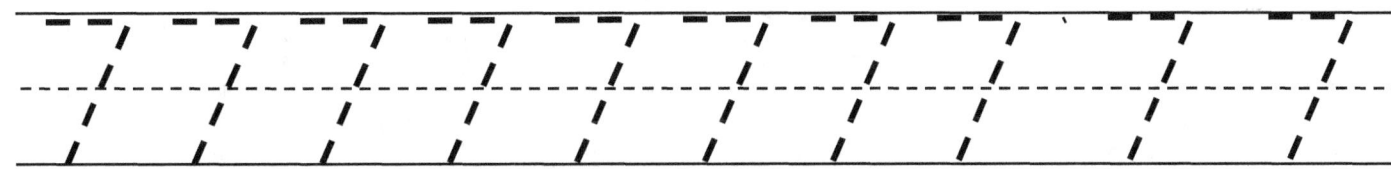

Count the calculator :

Circle the number 7:

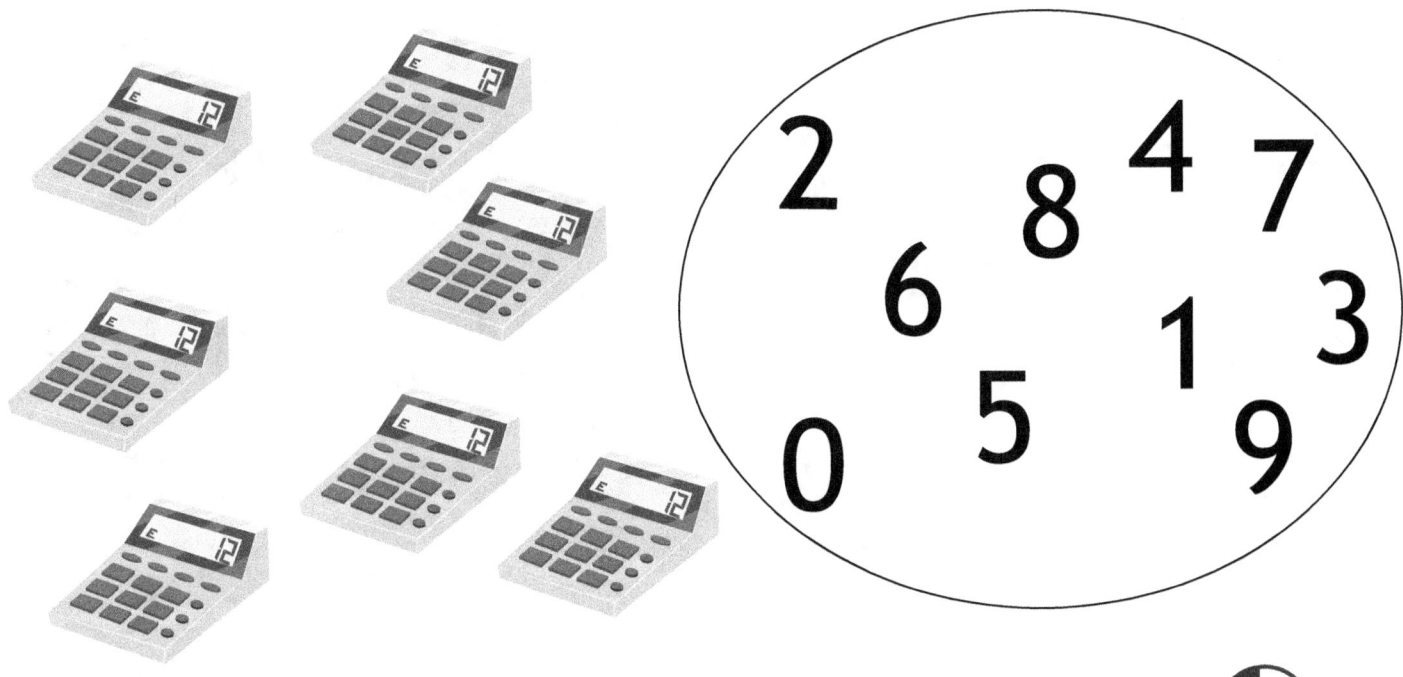

Tracing the number 8 (eight)

Kindergarten Numbers & Counting Worksheet

Practice tracing and printing the number 8 (eight).

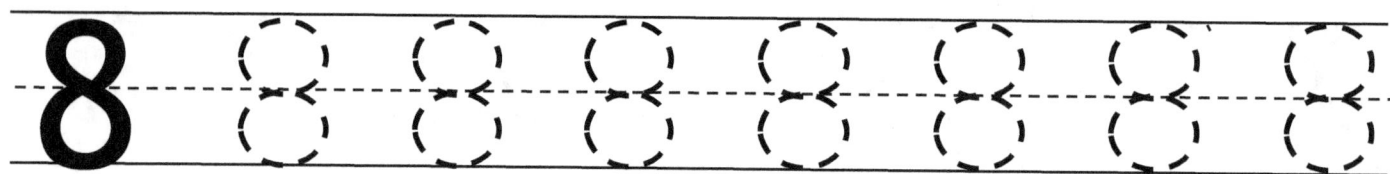

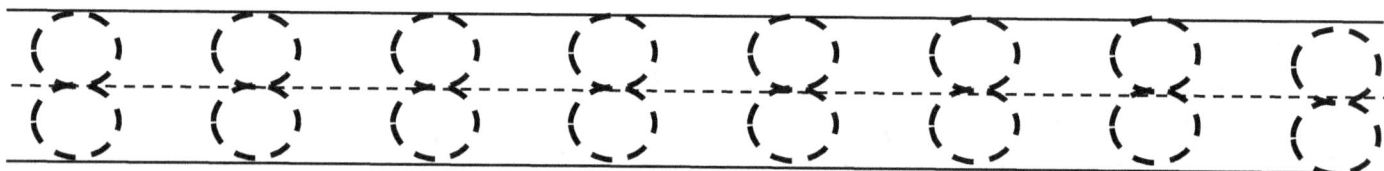

Tracing the number 9 (nine)
Kindergarten Numbers & Counting Worksheet

Practice tracing and printing the number 9 (nine).

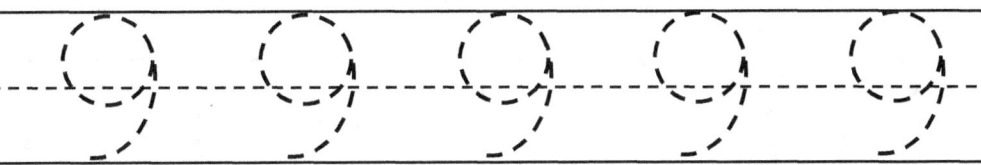

Count the pen:

Circle the number 9:

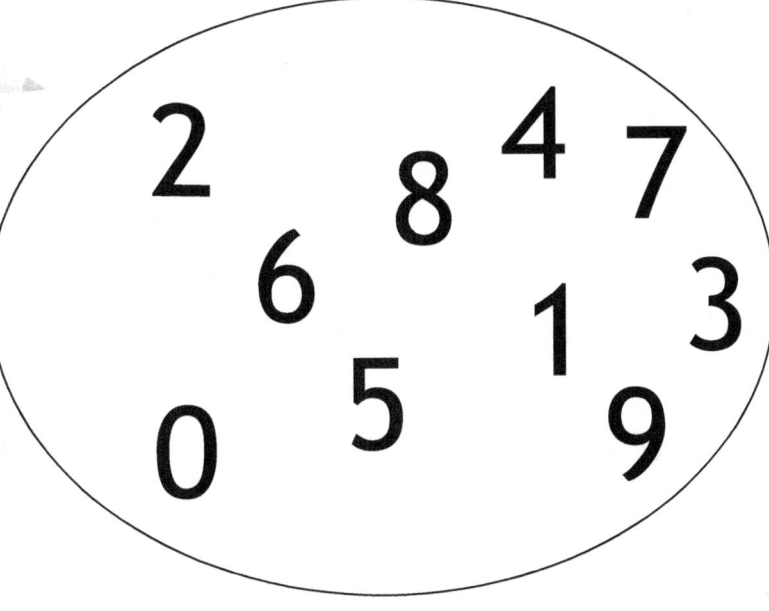

Tracing the number 10 (ten)

Kindergarten Numbers & Counting Worksheet

Practice tracing and printing the number 10 (ten).

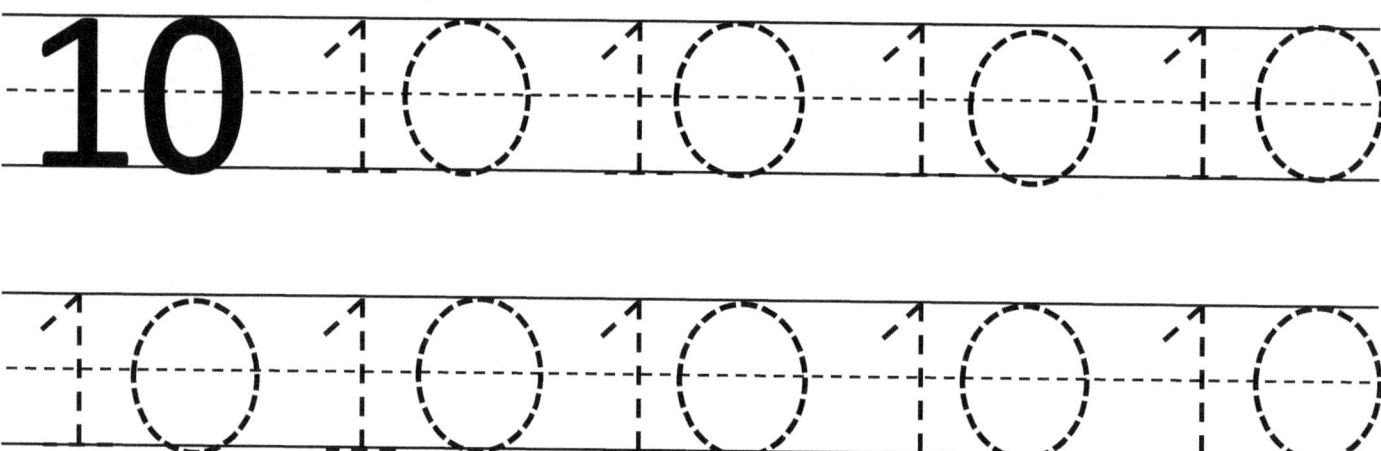

Count the alarm clock :

Circle the number 10:

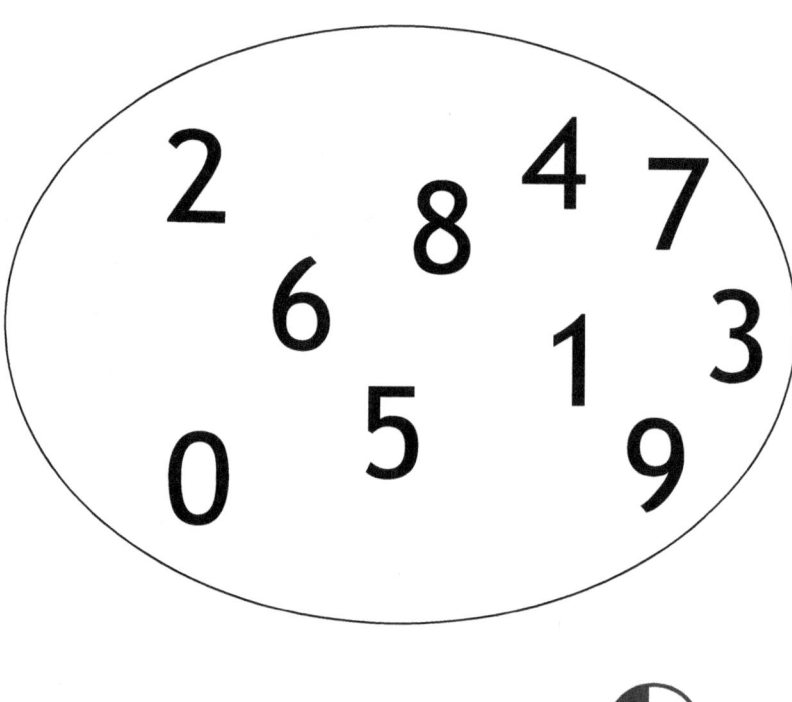

Counting Objects (numbers 1-10)

Grade 1 Counting Worksheet

Circle the correct number of objects:

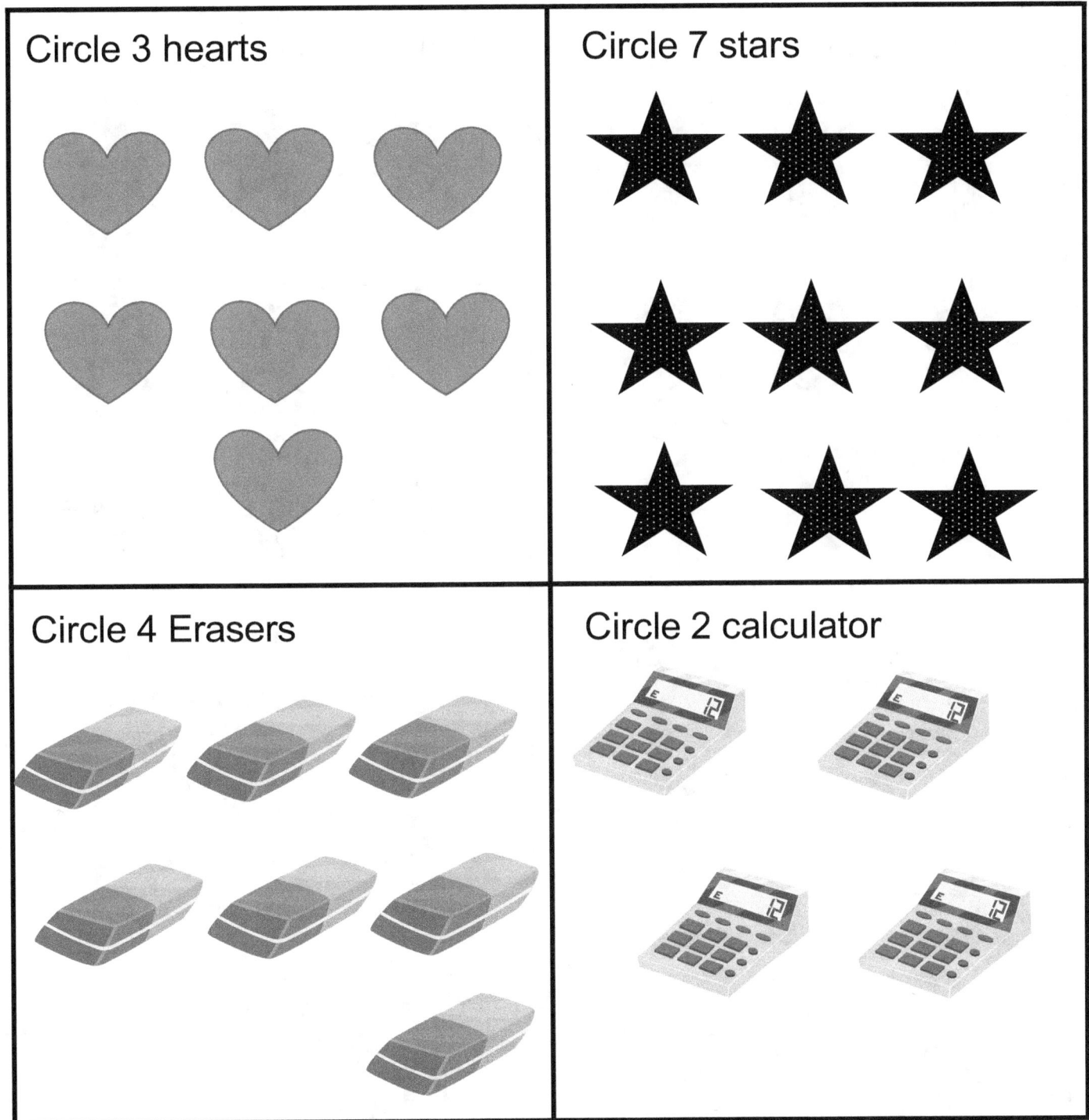

Counting Objects (numbers 1-10)

Grade 1 Counting Worksheet

Circle the correct number of objects:

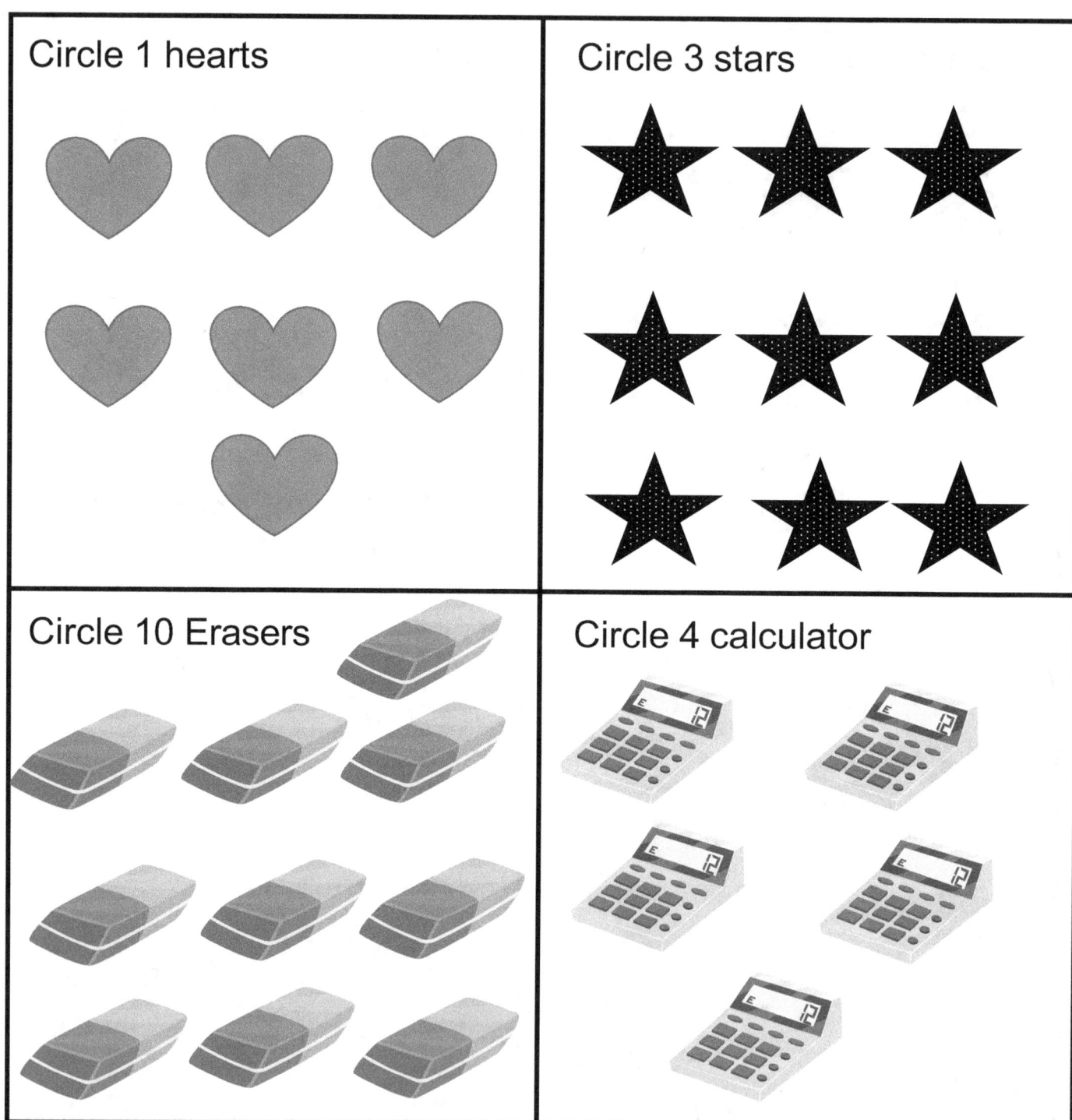

Counting Objects (numbers 1-10)

Grade 1 Counting Worksheet

Circle the correct number of objects:

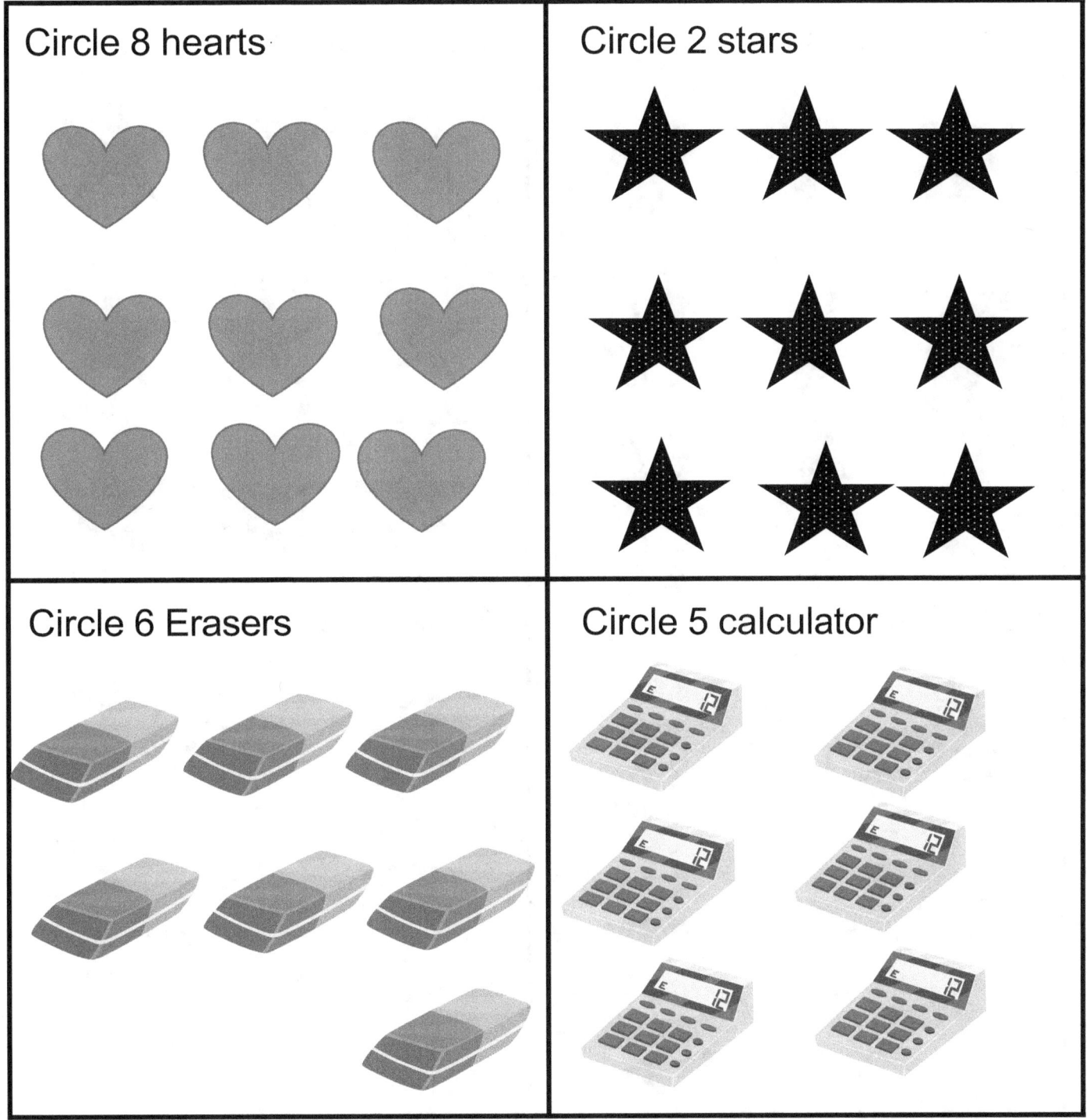

Counting Objects (numbers 1-10)

Grade 1 Counting Worksheet

Circle the correct number of objects:

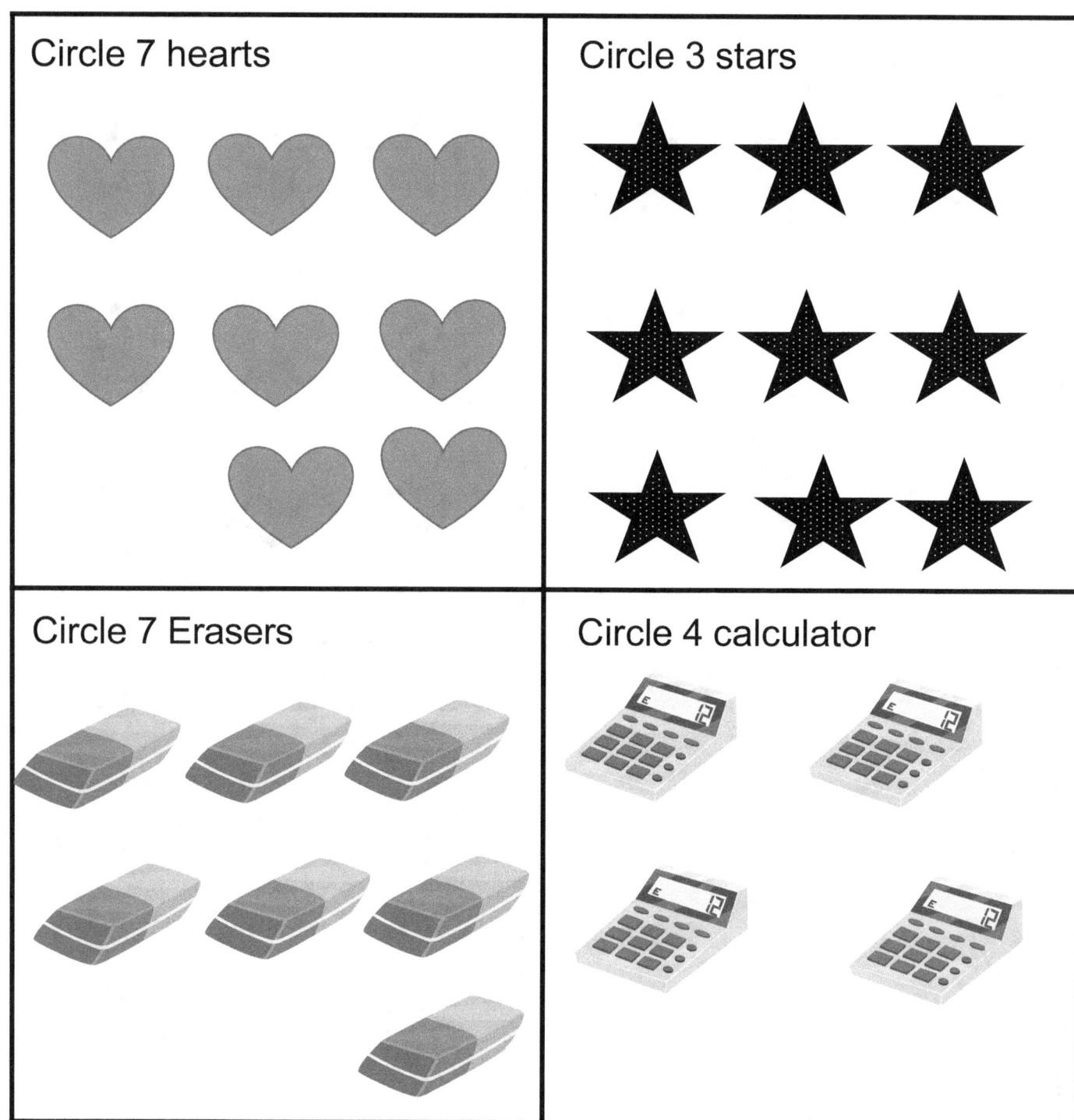

Counting Objects (numbers 1-10)

Grade 1 Counting Worksheet

Circle the correct number of objects:

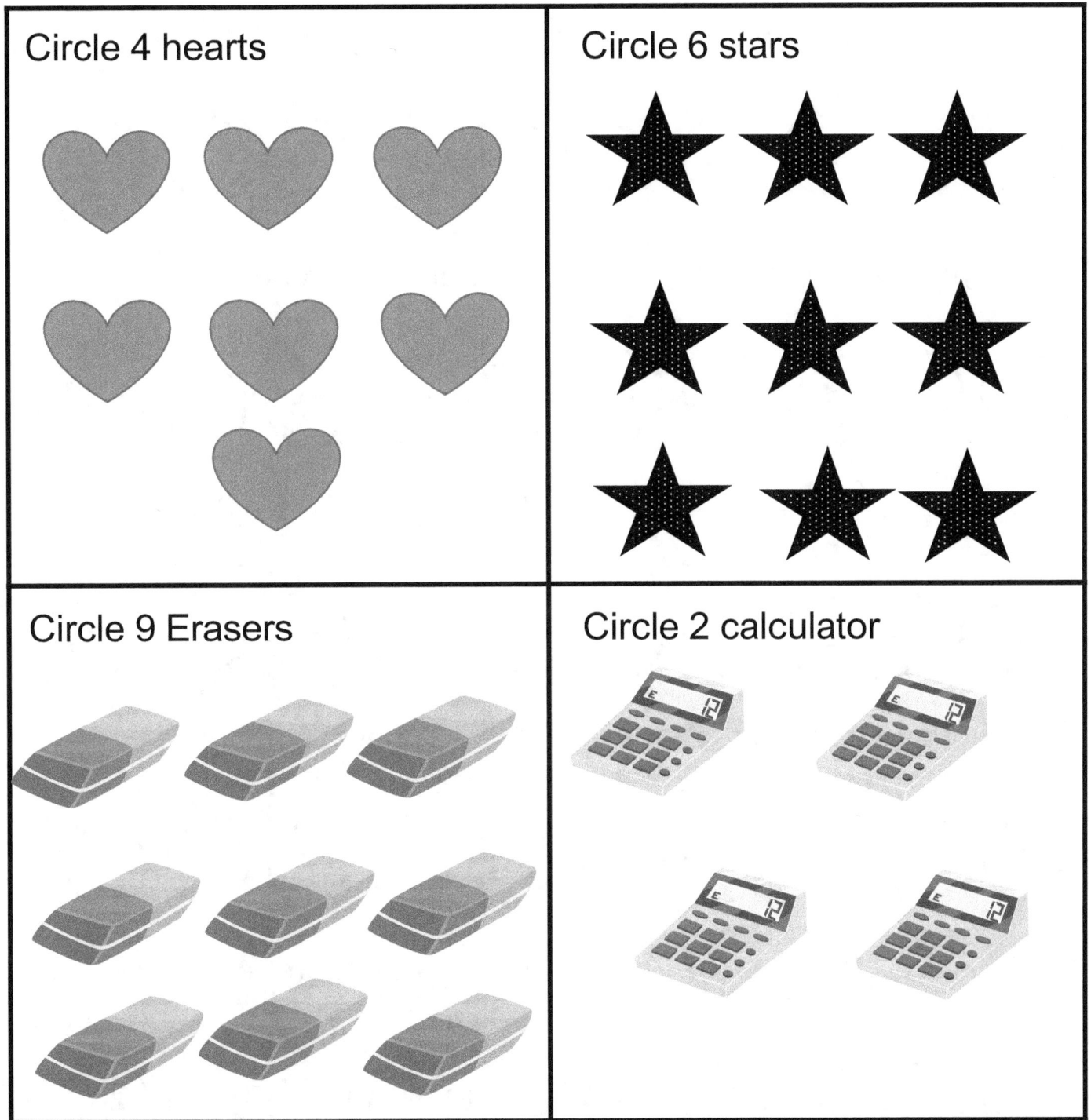

Counting by 2's
Grade 1 Counting Worksheet

1. Count by 2 from 5 to 15
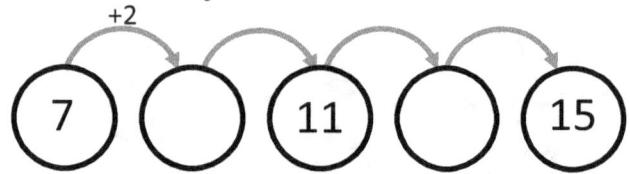

2. Count by 2 from 7 to 17
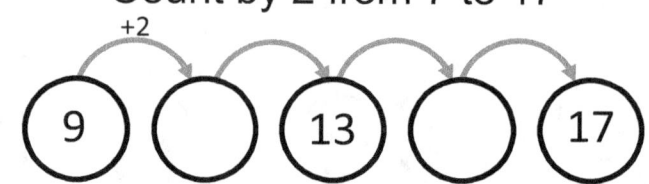

3. Count by 2 from 3 to 13
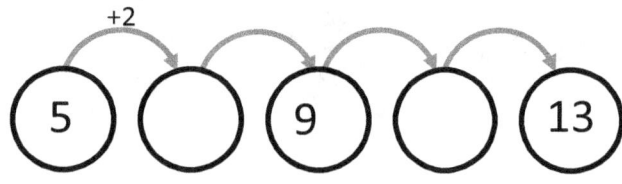

4. Count by 2 from 9 to 19
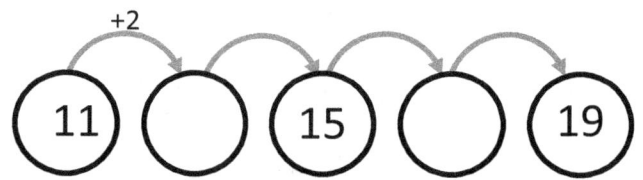

5. Count by 2 from 4 to 14
6. Count by 2 from 6 to 16
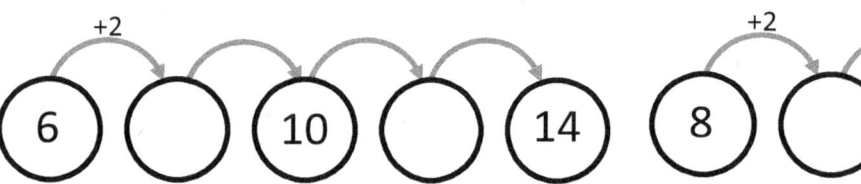
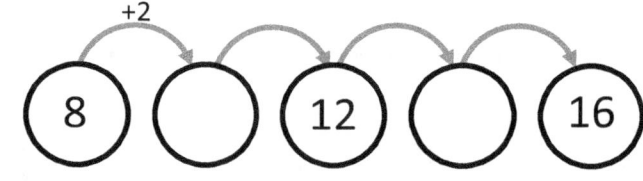

7. Count by 2 from 2 to 12   8. Count by 2 from 1 to 11
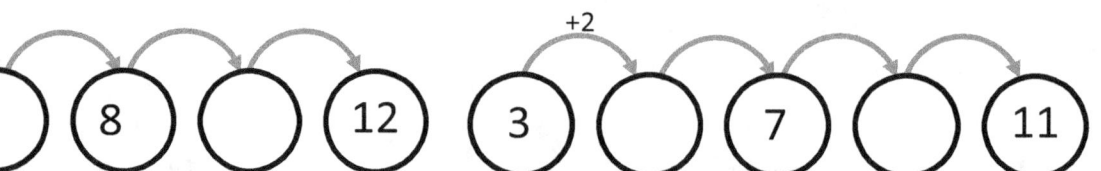

9. Count by 2 from 8 to 18   10. Count by 2 from 10 to 20

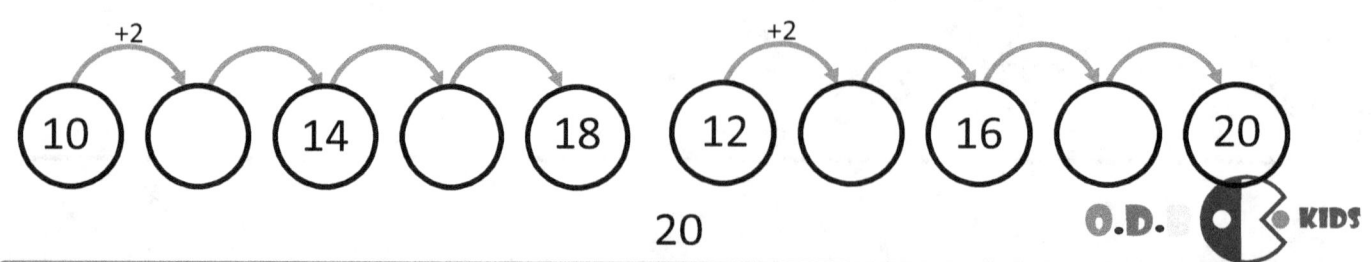

Counting by 2's
Grade 1 Counting Worksheet

1. Count by 2 from 7 to 17

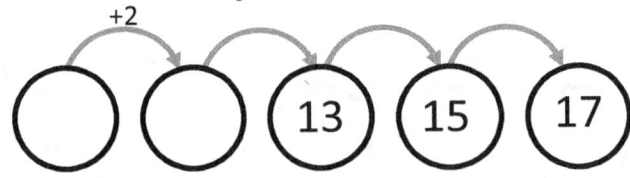

2. Count by 2 from 8 to 18

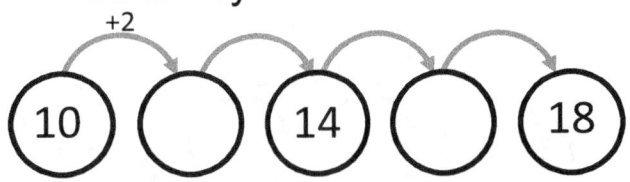

3. Count by 2 from 3 to 13

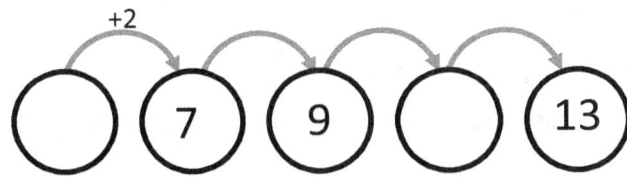

4. Count by 2 from 4 to 14

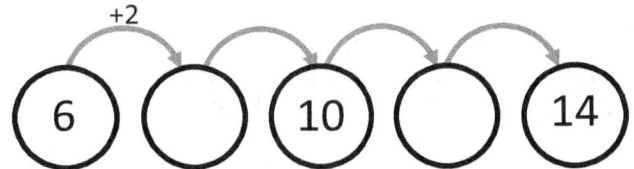

5. Count by 2 from 10 to 20

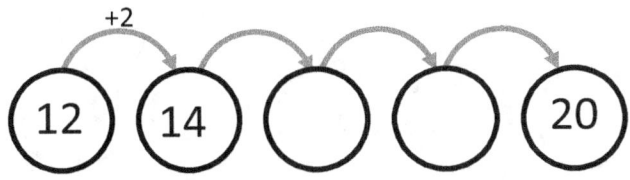

6. Count by 2 from 5 to 15

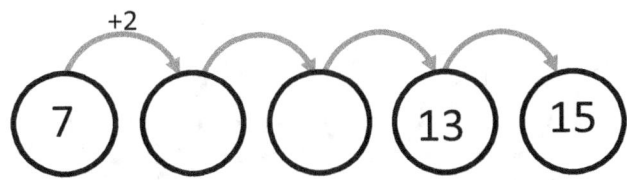

7. Count by 2 from 6 to 16

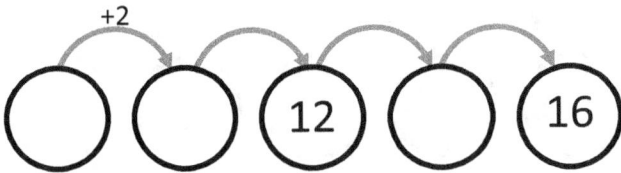

8. Count by 2 from 2 to 12

9. Count by 2 from 1 to 11

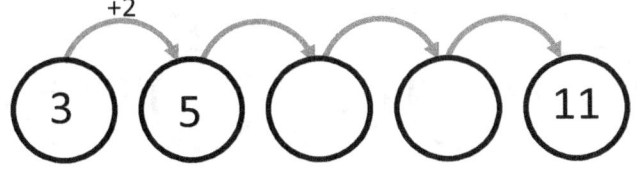

10. Count by 2 from 9 to 19

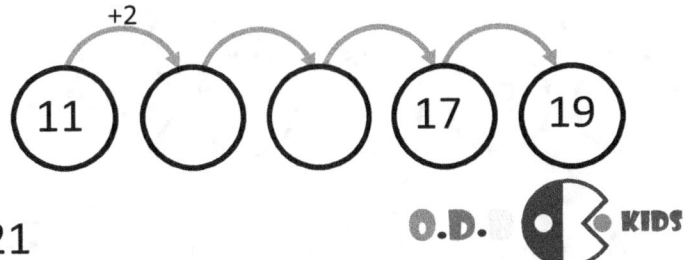

Counting by 2's
Grade 1 Counting Worksheet

9. Count by 2 from 8 to 18
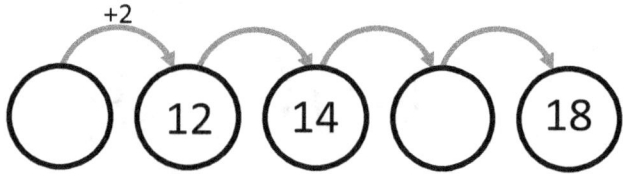

10. Count by 2 from 10 to 20
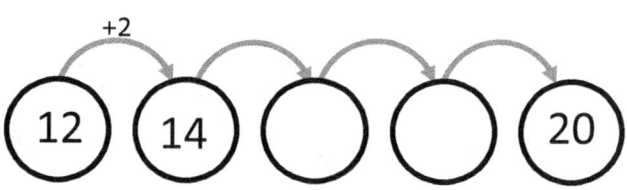

7. Count by 2 from 2 to 12
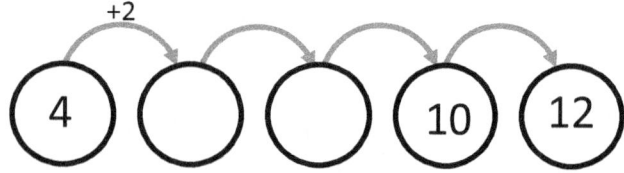

Count by 2 from 1 to 11
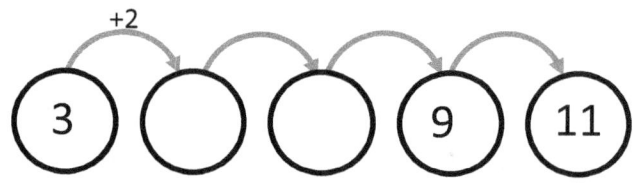

5. Count by 2 from 4 to 14
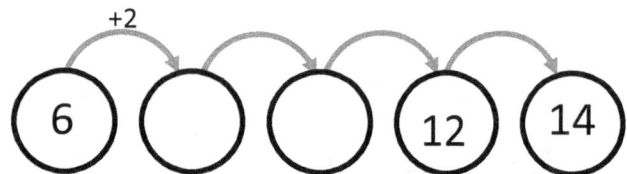

6. Count by 2 from 6 to 16
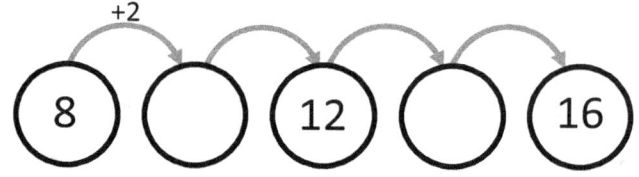

3. Count by 2 from 3 to 13
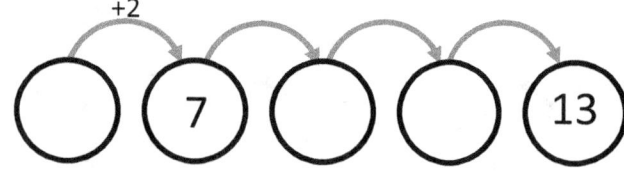

4. Count by 2 from 9 to 19
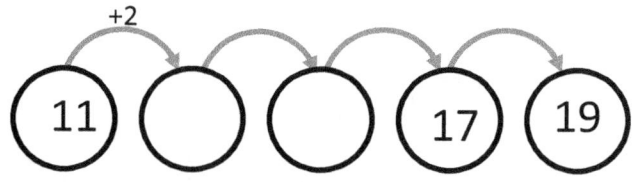

1. Count by 2 from 5 to 15
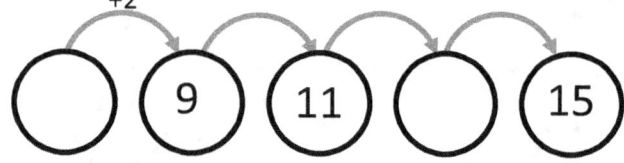

2. Count by 2 from 7 to 17

Counting by 2's
Grade 1 Counting Worksheet

5. Count by 2 from 10 to 20
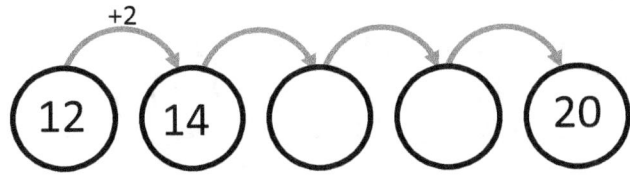

6. Count by 2 from 5 to 15
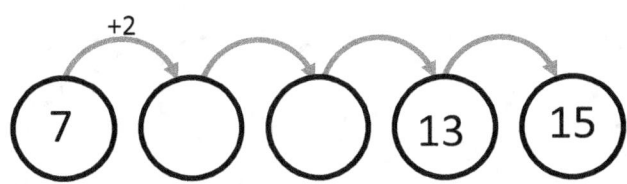

3. Count by 2 from 3 to 13

4. Count by 2 from 4 to 14

1. Count by 2 from 7 to 17

2. Count by 2 from 8 to 18
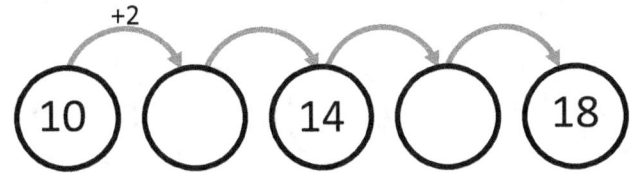

10. Count by 2 from 9 to 19
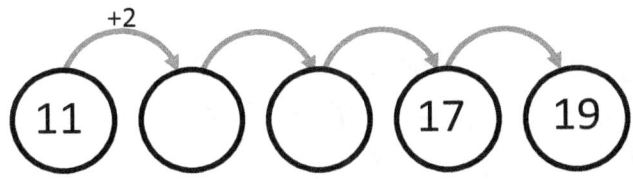

8. Count by 2 from 2 to 12

9. Count by 2 from 1 to 11
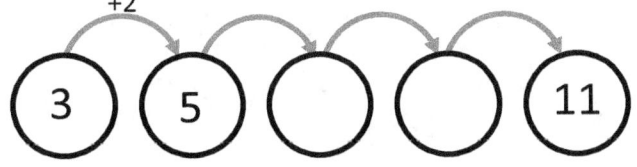

7. Count by 2 from 6 to 16
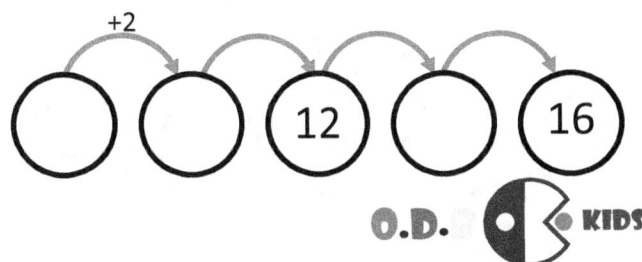

Counting by 2's
Grade 1 Counting Worksheet

1. Count by 2 from 5 to 15

2. Count by 2 from 7 to 17
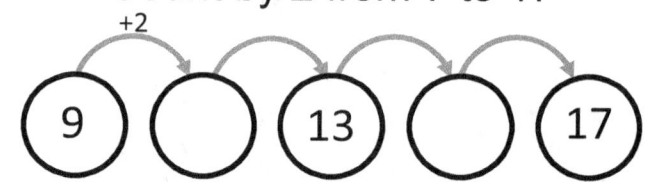

3. Count by 2 from 3 to 13
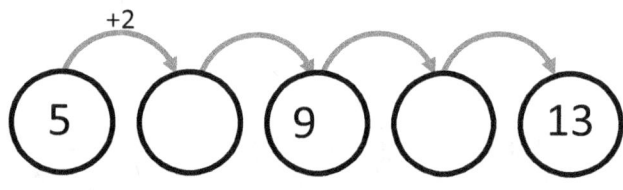

4. Count by 2 from 9 to 19
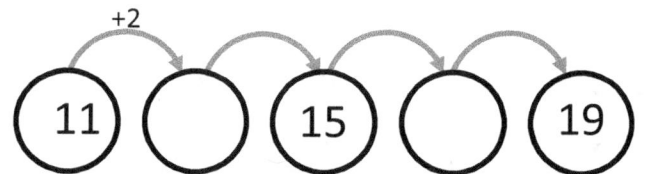

5. Count by 2 from 4 to 14

6. Count by 2 from 6 to 16
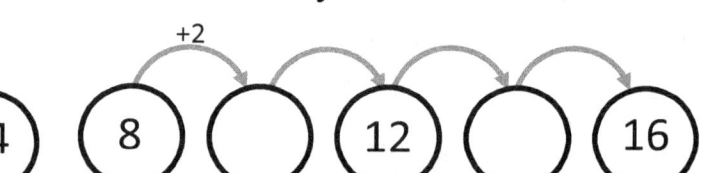

7. Count by 2 from 2 to 12

8. Count by 2 from 1 to 11

9. Count by 2 from 8 to 18

10. Count by 2 from 10 to 20
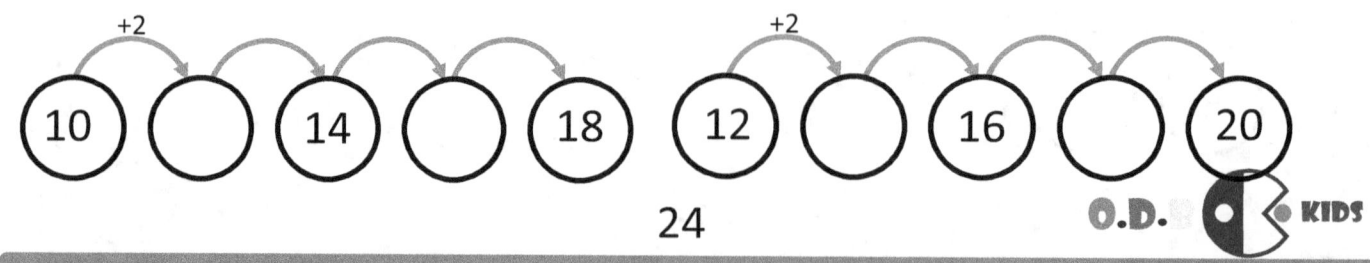

Counting by 2's
Grade 1 Counting Worksheet

1. Count by 2 from 7 to 17

2. Count by 2 from 8 to 18

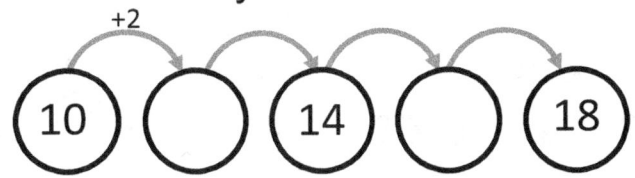

3. Count by 2 from 3 to 13

4. Count by 2 from 4 to 14

5. Count by 2 from 10 to 20

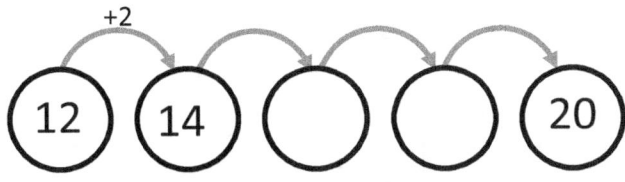

6. Count by 2 from 5 to 15

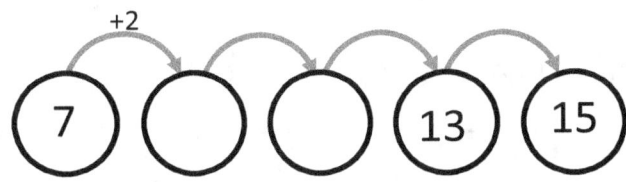

7. Count by 2 from 6 to 16

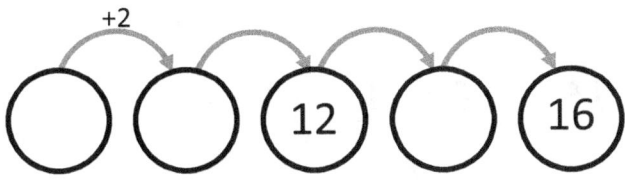

8. Count by 2 from 2 to 12

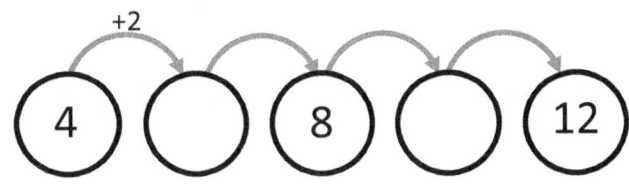

9. Count by 2 from 1 to 11

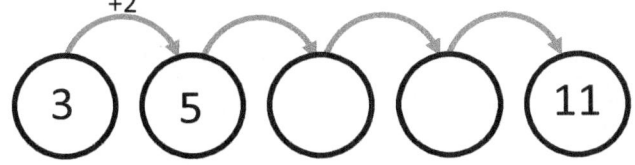

10. Count by 2 from 9 to 19

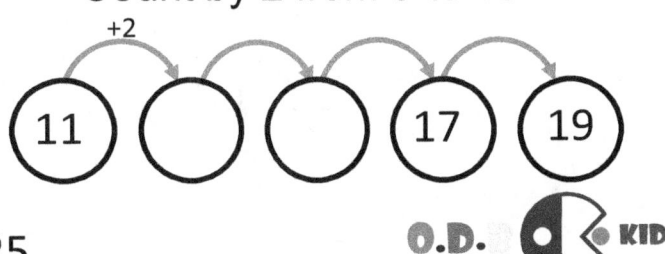

Identifying even / odd numbers 1-20

Grade 1 Counting & Numbers Worksheet

Circle the EVEN number(s).

1) 10 9 8 5 2 4

2) 9 10 7 18 20

3) 3 12 11 20 7 1

4) 2 5 17 13 11

5) 20 17 5 6 2 3

6) 10 8 3 2 1

7) 16 14 11 2 1 7

8) 14 7 5 18 19

9) 10 9 14 15 3

10) 2 13 16 12 6

Circle the ODD number(s).

11) 9 10 7 18 20

12) 3 10 2 8 9

13) 2 5 17 13 11

14) 15 9 5 17 7

15) 10 8 3 2 1

16) 14 16 9 12 5

17) 14 7 5 18 19

18) 20 10 11 8 9

19) 2 13 16 12 6

20) 13 17 6 2 20

Identifying even / odd numbers 1-20

Grade 1 Counting & Numbers Worksheet

Circle the EVEN number(s).

1) 10   17   8   5   3   4

2) 9   1   8   13   20

3) 3   12   11   20   7   8

4) 2   5   16   13   12

5) 7   16   5   6   2   3

6) 7   8   5   2   1

7) 18   15   11   3   16   7

8) 14   7   19   18   20

9) 13   9   14   15   4

10) 2   7   1   12   6   4

Circle the ODD number(s).

11) 9   5   7   1   20

12) 3   20   2   5   9

13) 2   5   17   13   15

14) 15   7   5   17   6

15) 6   8   3   2   17

16) 14   17   2   12   5

17) 14   7   8   18   19

18) 3   10   11   8   9

19) 2   13   16   19   6

20) 9   17   6   2   20

Identifying even / odd numbers 1-20

Grade 1 Counting & Numbers Worksheet

Circle the EVEN number(s).

1) 17   19   15   1

2) 9   8   15   1

3) 12   5   3   10

4) 18   15   11   5

5) 12   6   1   19

6) 20   10   7   5

7) 11   17   14   8

8) 1   19   2   20

9) 3   6   15   5

10) 15   5   20   8

Circle the ODD number(s).

11) 1   19   16   6

12) 19   9   2   3

13) 16   14   13   12

14) 6   11   17   13

15) 18   4   19   9

16) 14   8   20   6

17) 8   4   12   18

18) 17   2   15   11

19) 7   10   15   2

20) 7   9   20   2

Identifying even / odd numbers 1-20

Grade 1 Counting & Numbers Worksheet

Circle the EVEN number(s).

1) 5 1 82 50

2) 48 47 4 86

3) 60 6 52 53

4) 6 4 2 7

5) 19 6 7 66

6) 85 45 97 6

7) 50 4 2 74

8) 2 8 60 62

9) 6 2 3 62

10) 41 52 7 92

Circle the ODD number(s).

11) 91 78 26 9

12) 35 1 7 64

13) 53 39 38 94

14) 32 80 91 57

15) 2 31 97 68

16) 55 7 17 8

17) 96 92 68 1

18) 4 6 38 8

19) 58 96 27 5

20) 17 100 1 3

Identifying even / odd numbers 1-20

Grade 1 Counting & Numbers Worksheet

Circle the EVEN number(s).

1) 64   587   478   52
2) 61   35   303   247
3) 3   988   261   1
4) 665   930   318   268
5) 9   104   2   914
6) 437   2   41   348
7) 960   474   5   20
8) 960   266   456   55
9) 2   8   62   898
10) 898   1   9   7

Circle the ODD number(s).

11) 6   687   181   880
12) 299   36   53   988
13) 7   815   416   807
14) 350   36   687   152
15) 98   22   981   5
16) 50   7   523   659
17) 23   397   188   567
18) 759   76   32   837
19) 20   3   5   17
20) 43   548   29   2

Number Chart from 1 to 100 (half-full)
Grade 1 Number Charts

Count by 1 from 1 to 100

1	2	3	4	5	6	7		9	
				15				19	
	22	23		25	26	27	28		
	32	33		35	36		38	39	
		43		45				49	50
51				55		57		59	60
	62		64		66			69	
	72				76	77			
	82		84						90
91	92				96	97		99	100

Number Chart from 1 to 100 (half-full)

Grade 1 Number Charts

Count by 1 from 1 to 100

1	2	3			6		8	9	10
		13	14		16	17	18		20
	22		24	25		27	28		
31	32	33		35	36	37			40
	42		44			47	48	49	50
				55		57	58		
		63		65		67	68		
				75			78	79	
		83	84	85	86			89	90
91	92		94			97	98	99	100

Number Chart from 1 to 100 (half-full)

Grade 1 Number Charts

Count by 1 from 1 to 100

1	2	3	4			7		9	10
11			14			17		19	
	22					27	28		30
31	32	33			36		38		
	42			45	46		48	49	
51	52	53			56		58		
61			64	65			68		70
	72	73			76				
81		83			86		88	89	90
	92		94		96	97		99	100

Number Chart from 1 to 100 (half-full)
Grade 1 Number Charts

Count by 1 from 1 to 100

1	2				6				
		13		15					20
				25		27			
	32	33				37		39	40
				45					
	52					57			
61				65		67			
71						76			80
						86			
	92			94	95				100

Numbers as words (0-20)

Grade 1 Numbers Worksheet

Circle the correct number for each word.

eight	5	13	8
sixteen	16	6	19
fourteen	14	24	4
twenty	2	12	20
ten	9	10	2
three	3	6	9
thirteen	16	13	4
nineteen	9	16	19
eleven	11	12	1
twelve	12	11	1

Numbers as words (1-20)

Grade 1 Numbers Worksheet

Draw a line between the number to its word.

14	fourteen
8	seventeen
11	thirteen
20	ten
6	three
10	eight
17	eleven
19	six
3	nineteen
13	twenty

Numbers as words (0-30)
Grade 1 Numbers Worksheet

Circle the correct number for each word.

twenty-four	14	24	4
twenty	20	2	16
thirty	13	21	30
twelve	10	11	12
sixteen	16	13	19
twenty-nine	23	26	29
eight	16	8	4
fourteen	24	16	14
twenty-one	22	12	21
twenty-seven	7	20	27

Numbers as words (1-30)

Grade 1 Numbers Worksheet

Draw a line between the number to its word.

2	twenty-one
12	thirty
22	eleven
27	fifteen
15	twenty-four
11	twelve
21	twenty-two
30	twenty-seven
29	two
24	twenty-nine

Numbers as words (0-120)
Grade 1 Numbers Worksheet

Circle the correct number for each word.

eighty-four	44	88	84
fifty-six	56	65	66
thirty-three	13	23	33
one hundred	101	100	10
fifteen	15	55	25
ninety-nine	66	96	99
eighty	68	80	40
one hundred twelve	112	61	114
twenty-one	29	12	21
sixty-seven	61	68	67

Numbers as words (1-120)

Grade 1 Numbers Worksheet

Draw a line between the number to its word.

50	fourteen
58	one hundred fourteen
6	eighty-five
100	fifty
14	thirty-three
114	six
85	seventy-four
33	twenty-one
74	fifty-eight
21	one hundred

Counting Objects (up to 20)

Grade 1 Counting Worksheet

Count the objects and write the number in the box.

 = ☐

 = ☐

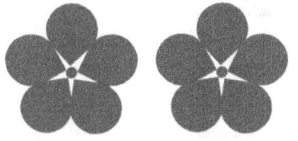 = ☐

 = ☐

  = ☐

Counting Objects (up to 20)

Grade 1 Counting Worksheet

Count the objects and write the number in the box.

☺ ☺ ☺ ☺ ☺ ☺ ☺ ☺ = ☐

 = ☐

☁ ☁ ☁ ☁ ☁ = ☐

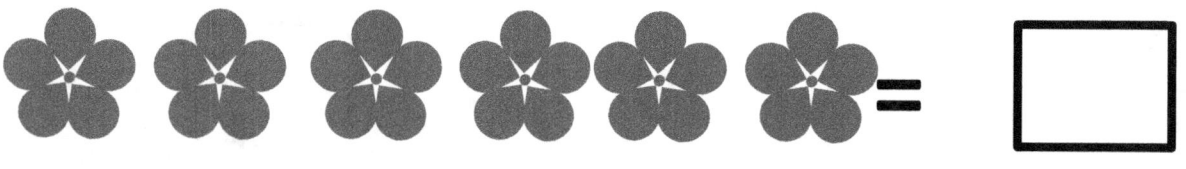 = ☐

♥ ♥ ♥ ♥ ♥ ♥
♥ ♥ ♥ = ☐

Counting Objects (up to 20)

Grade 1 Counting Worksheet

Count the objects and write the number in the box.

 = ☐

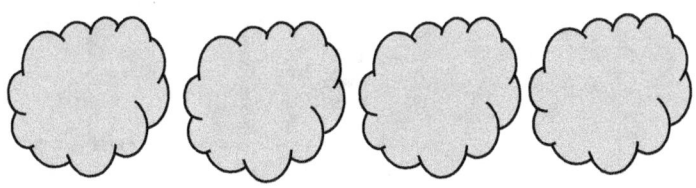 = ☐

 = ☐

 = ☐

 = ☐

Counting Objects (up to 20)

Grade 1 Counting Worksheet

Count the objects and write the number in the box.

 =

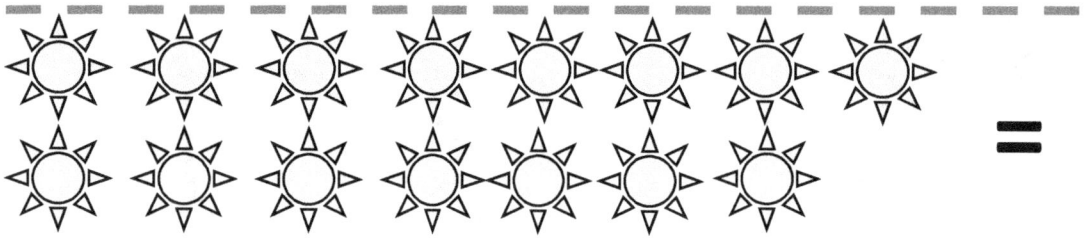 =

 = □

 = □

 =

Counting Objects (up to 20)

Grade 1 Counting Worksheet

Count the objects and write the number in the box.

 =

 =

 =

 =

 =

Counting practice - before / after (1-20)

Grade 1 Counting Worksheet

Write in the missing numbers.

1: | 5 | |

2: | 9 | |

3: | | 7 |

4: | 2 | |

5: | | 4 |

6: | 15 | |

7: | 17 | |

8: | 12 | |

9: | | 8 |

10: | 8 | | 10 |

11: | 14 | |

12: | 6 | |

13: | | 3 |

14: | | 10 |

15: | 15 | | 17 |

16: | | 16 |

17: | | 1 |

18: | 3 | |

Counting practice - before / after (1-20)

Grade 1 Counting Worksheet

Write in the missing numbers.

1: | 6 | | 2: | 10 | | 3: | | 6 |

4: | 3 | | 5: | | 5 | 6: | 14 | |

7: | 18 | | 8: | 13 | | 9: | | 9 |

10: | 9 | | 11 | 11: | 1 | | 12: | 11 | |

13: | | 15 | 14: | | 20 | 15: | 14 | | 16 |

16: | | 19 | 17: | | 8 | 18: | 2 | |

Counting practice - before / after (1-20)

Grade 1 Counting Worksheet

Write in the missing numbers.

1: | 3 | ___ | 2: | 12 | ___ | 3: | ___ | 14 |

4: | ___ | 4 | 5: | 9 | ___ | 6: | ___ | 7 |

7: | 17 | ___ | 19 | 8: | ___ | 2 | 9: | 7 | ___ |

10: | 11 | ___ | 11: | 17 | ___ | ___ | 12: | 3 | ___ |

13: | ___ | 13 | 14: | ___ | 19 | 15: | 5 | ___ | 7 |

16: | ___ | 9 | 17: | ___ | 10 | 18: | 2 | ___ |

Counting practice - before / after (1-20)

Grade 1 Counting Worksheet

Write in the missing numbers.

1: | 3 | |

2: | 12 | |

3: | | 14 |

4: | | 4 |

5: | 9 | |

6: | | 7 |

7: | 17 | | 19 |

8: | | 2 |

9: | 7 | |

10: | 11 | |

11: | 17 | | |

12: | 3 | |

13: | | 13 |

14: | | 19 |

15: | 5 | | 7 |

16: | | 9 |

17: | | 10 |

18: | 2 | |

Identifying tens and ones

Place Value

Fill in the correct tens and ones for the given numbers.

tens [ ] and ones [ ] = 86

tens [ ] and ones [ ] = 16

tens [ ] and ones [ ] = 36

tens [ ] and ones [ ] = 25

tens [ ] and ones [ ] = 76

tens [ ] and ones [ ] = 14

tens [ ] and ones [ ] = 63

tens [ ] and ones [ ] = 17

tens [ ] and ones [ ] = 23

Identifying tens and ones

Place Value

Fill in the correct tens and ones for the given numbers.

tens		and	ones		= 30
tens		and	ones		= 25
tens		and	ones		= 46
tens		and	ones		= 70
tens		and	ones		= 89
tens		and	ones		= 73
tens		and	ones		= 19
tens		and	ones		= 37
tens		and	ones		= 94

Identifying tens and ones

Place Value

Fill in the correct tens and ones for the given numbers.

tens [ ] and ones [ ] = 42

tens [ ] and ones [ ] = 67

tens [ ] and ones [ ] = 13

tens [ ] and ones [ ] = 93

tens [ ] and ones [ ] = 64

tens [ ] and ones [ ] = 57

tens [ ] and ones [ ] = 72

tens [ ] and ones [ ] = 16

tens [ ] and ones [ ] = 92

Identifying tens and ones

Place Value

Fill in the correct tens and ones for the given numbers.

tens		and	ones		= 78
tens		and	ones		= 29
tens		and	ones		= 37
tens		and	ones		= 63
tens		and	ones		= 76
tens		and	ones		= 94
tens		and	ones		= 17
tens		and	ones		= 38
tens		and	ones		= 18

Identifying tens and ones

Place Value

Fill in the correct tens and ones for the given numbers.

tens		and	ones		= 91
tens		and	ones		= 26
tens		and	ones		= 37
tens		and	ones		= 12
tens		and	ones		= 88
tens		and	ones		= 97
tens		and	ones		= 34
tens		and	ones		= 50
tens		and	ones		= 57

Identifying tens and ones

Place Value

Fill in the correct tens and ones for the given numbers.

tens		and	ones		= 15
tens		and	ones		= 67
tens		and	ones		= 94
tens		and	ones		= 36
tens		and	ones		= 43
tens		and	ones		= 49
tens		and	ones		= 84
tens		and	ones		= 22
tens		and	ones		= 33

Combining tens and ones

Place Value Worksheet

Fill in the correct tens and ones for the given numbers.

☐ = 2 tens and 4 ones

☐ = 1 ten and 2 ones

☐ = 6 tens and 7 ones

☐ = 9 tens and 6 ones

☐ = 5 tens and 5 ones

☐ = 8 tens and 0 ones

☐ = 3 tens and 9 ones

☐ = 4 tens and 8 ones

■ an example of

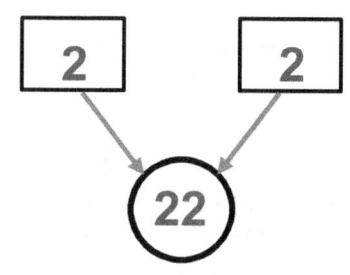
2 tens and 2 ones

Combining tens and ones

Place Value Worksheet

Fill in the correct tens and ones for the given numbers.

☐ = 5 tens and 4 ones

☐ = 2 tens and 3 ones

☐ = 6 tens and 7 ones

☐ = 8 tens and 1 one

☐ = 3 tens and 6 ones

☐ = 7 tens and 5 ones

☐ = 4 tens and 2 ones

☐ = 1 ten and 0 ones

Combining tens and ones

Place Value Worksheet

Fill in the correct tens and ones for the given numbers.

	=	8 tens and 3 ones
	=	7 tens and 8 ones
	=	1 ten and 4 ones
	=	3 tens and 3 ones
	=	4 tens and 1 one
	=	6 tens and 0 ones
	=	9 tens and 6 ones
	=	2 tens and 5 ones

Combining tens and ones

Place Value Worksheet

Fill in the correct tens and ones for the given numbers.

☐	=	2 tens and 0 ones
☐	=	7 tens and 9 ones
☐	=	5 tens and 1 one
☐	=	4 tens and 7 ones
☐	=	6 tens and 3 ones
☐	=	3 tens and 8 ones
☐	=	8 tens and 6 ones
☐	=	1 ten and 4 ones

Combining tens and ones

Place Value Worksheet

Fill in the correct tens and ones for the given numbers.

☐ = 3 tens and 7 ones

☐ = 2 ten and 8 ones

☐ = 4 tens and 0 ones

☐ = 9 tens and 1 one

☐ = 7 tens and 5 ones

☐ = 6 tens and 4 ones

☐ = 5 tens and 4 ones

☐ = 8 tens and 7 ones

Combining tens and ones

Place Value Worksheet

Fill in the correct tens and ones for the given numbers.

☐ = 5 tens and 8 ones

☐ = 6 tens and 6 ones

☐ = 1 ten and 9 ones

☐ = 3 tens and 7 ones

☐ = 4 tens and 3 ones

☐ = 8 tens and 7 ones

☐ = 9 tens and 4 ones

☐ = 2 tens and 6 ones

Counting by 2's
Grade 1 Counting Worksheet

1. Count by 2 from 5 to 15

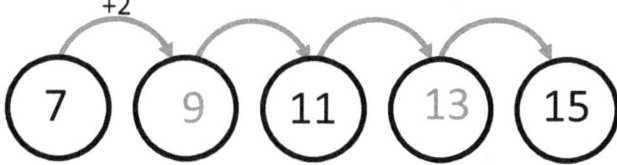

2. Count by 2 from 7 to 17

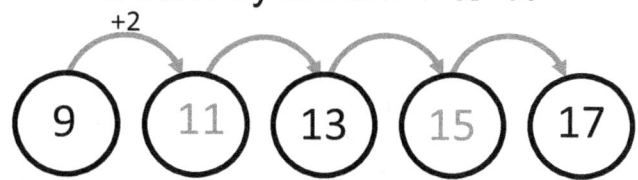

3. Count by 2 from 3 to 13

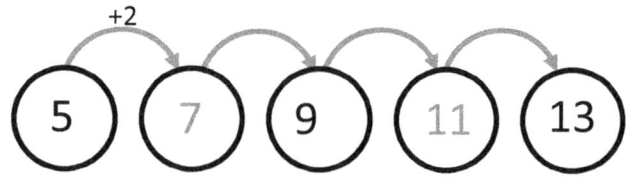

4. Count by 2 from 9 to 19

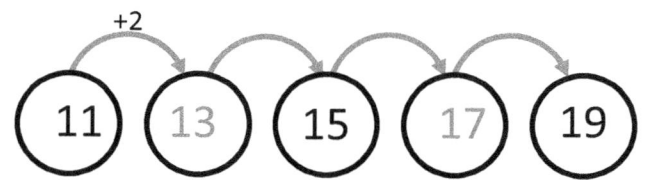

5. Count by 2 from 4 to 14   6. Count by 2 from 6 to 16

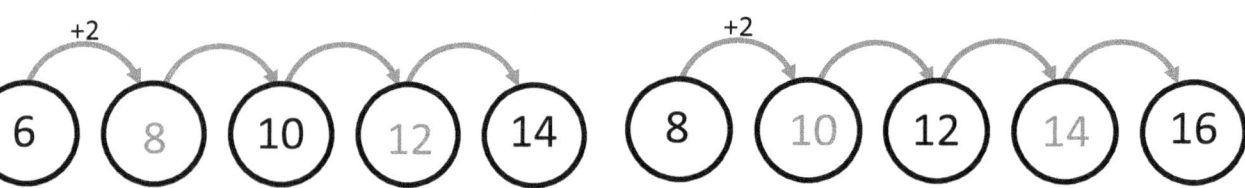

7. Count by 2 from 2 to 12   8. Count by 2 from 1 to 11

9. Count by 2 from 8 to 18   10. Count by 2 from 10 to 20

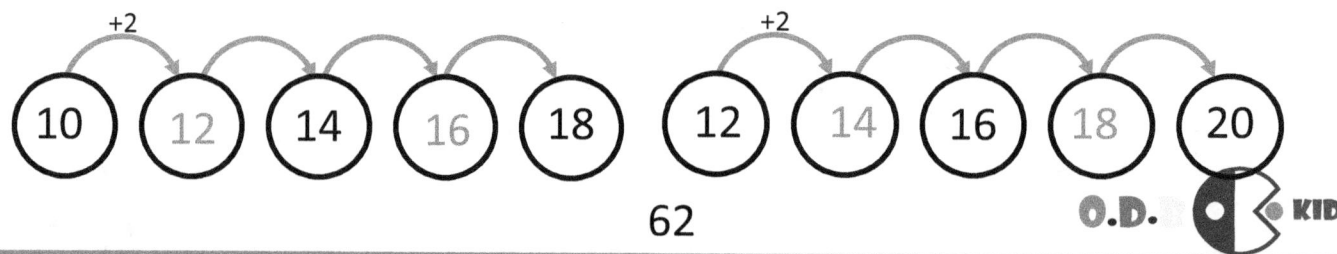

Counting by 2's
Grade 1 Counting Worksheet

1. Count by 2 from 7 to 17

2. Count by 2 from 8 to 18

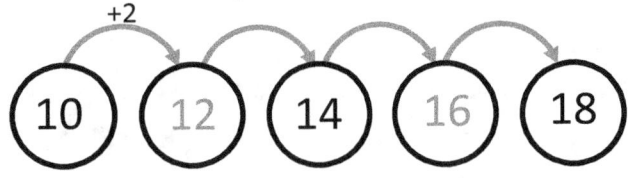

3. Count by 2 from 3 to 13

4. Count by 2 from 4 to 14

5. Count by 2 from 10 to 20

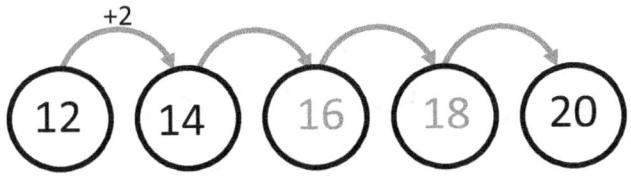

6. Count by 2 from 5 to 15

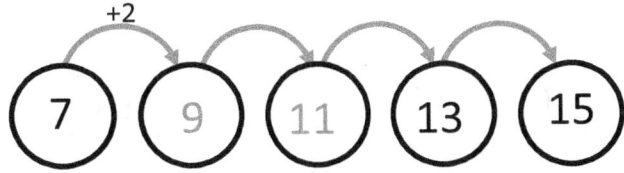

7. Count by 2 from 6 to 16

8. Count by 2 from 2 to 12

9. Count by 2 from 1 to 11

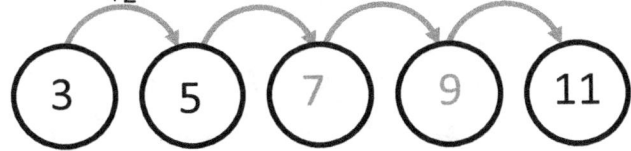

10. Count by 2 from 9 to 19

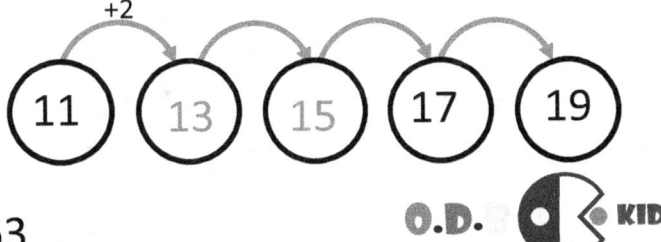

Counting by 2's
Grade 1 Counting Worksheet

9. Count by 2 from 8 to 18

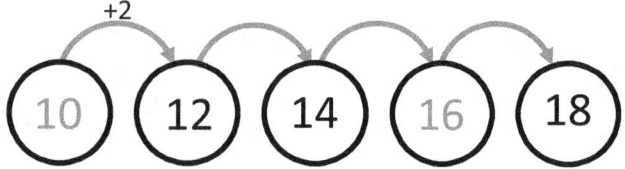

10. Count by 2 from 10 to 20

7. Count by 2 from 2 to 12

Count by 2 from 1 to 11

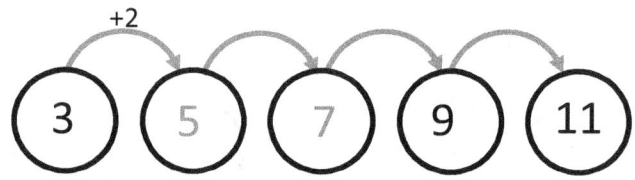

5. Count by 2 from 4 to 14

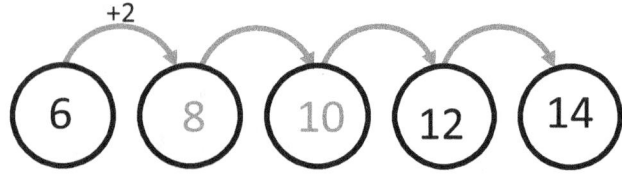

6. Count by 2 from 6 to 16

3. Count by 2 from 3 to 13

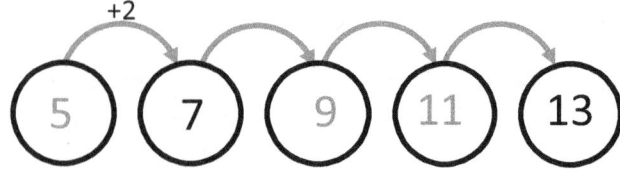

4. Count by 2 from 9 to 19

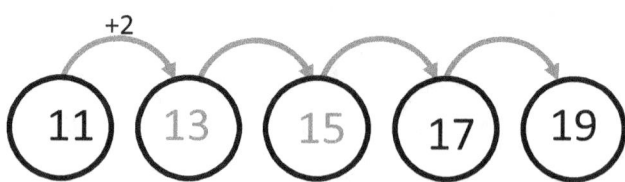

1. Count by 2 from 5 to 15

2. Count by 2 from 7 to 17

Counting by 2's
Grade 1 Counting Worksheet

5. Count by 2 from 10 to 20
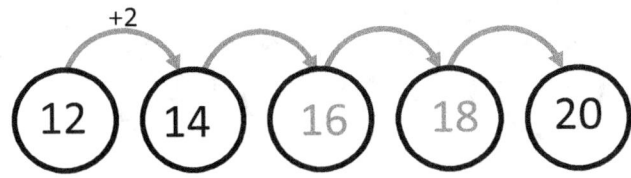

6. Count by 2 from 5 to 15
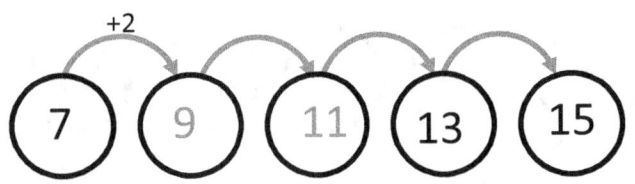

3. Count by 2 from 3 to 13
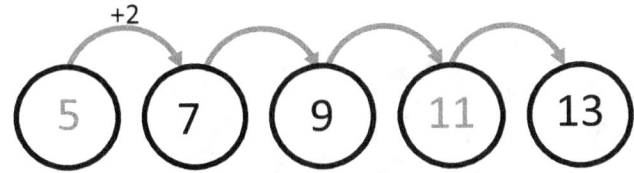

4. Count by 2 from 4 to 14
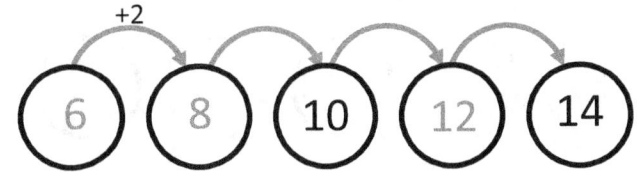

1. Count by 2 from 7 to 17

2. Count by 2 from 8 to 18
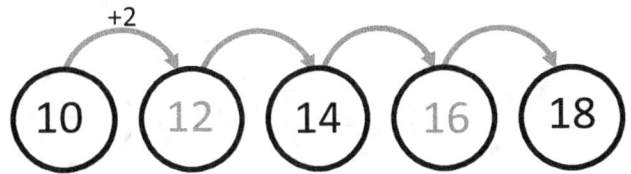

10. Count by 2 from 9 to 19

8. Count by 2 from 2 to 12
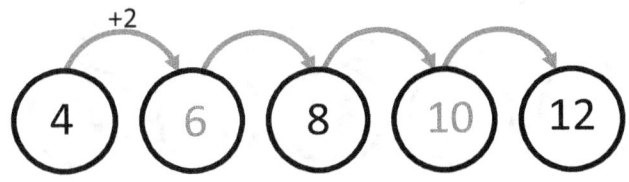

9. Count by 2 from 1 to 11

7. Count by 2 from 6 to 16
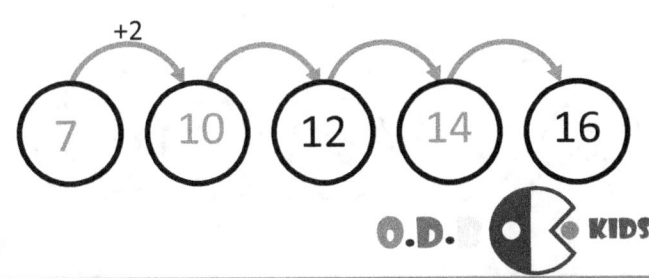

Counting by 2's
Grade 1 Counting Worksheet

1. Count by 2 from 5 to 15

2. Count by 2 from 7 to 17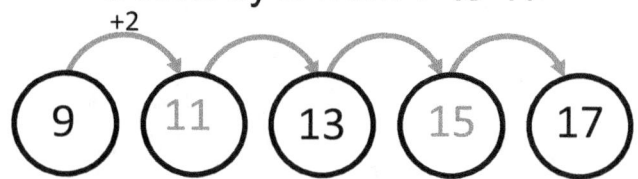

3. Count by 2 from 3 to 13

4. Count by 2 from 9 to 19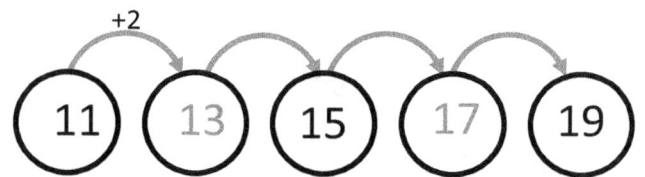

5. Count by 2 from 4 to 14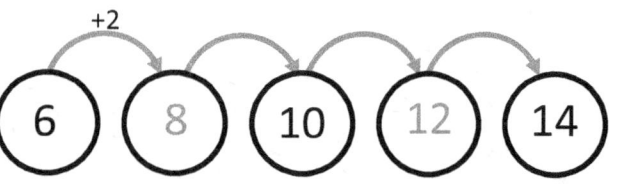

6. Count by 2 from 6 to 16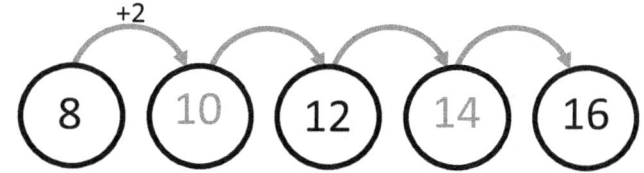

7. Count by 2 from 2 to 12

8. Count by 2 from 1 to 11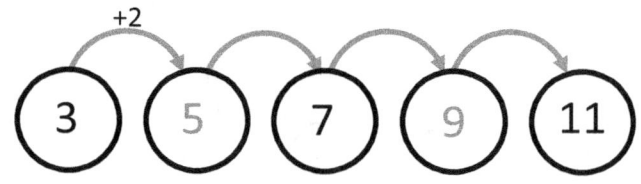

9. Count by 2 from 8 to 18

10. Count by 2 from 10 to 20

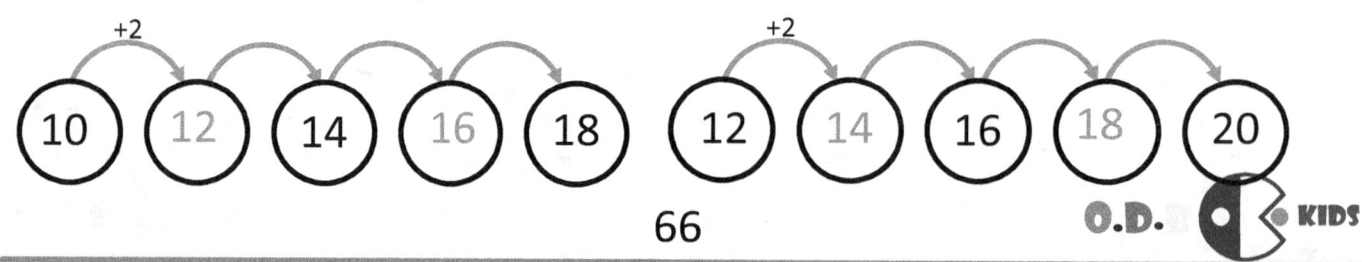

Counting by 2's
Grade 1 Counting Worksheet

1. Count by 2 from 7 to 17

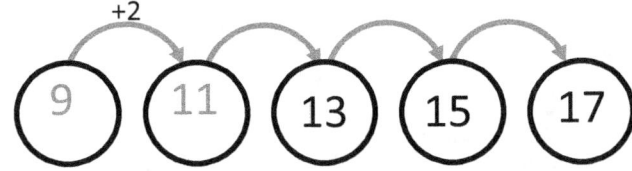

2. Count by 2 from 8 to 18

3. Count by 2 from 3 to 13

4. Count by 2 from 4 to 14

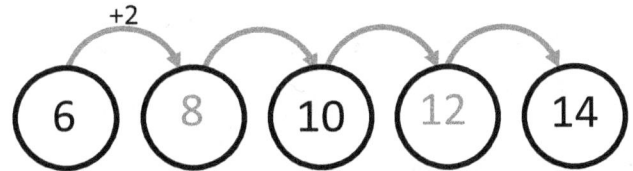

5. Count by 2 from 10 to 20

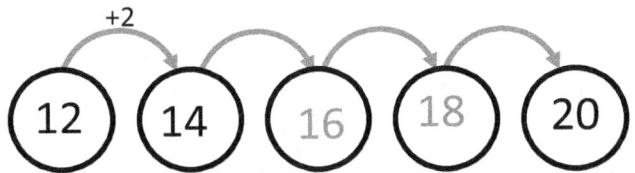

6. Count by 2 from 5 to 15

7. Count by 2 from 6 to 16

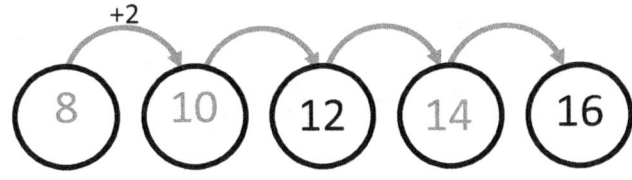

8. Count by 2 from 2 to 12

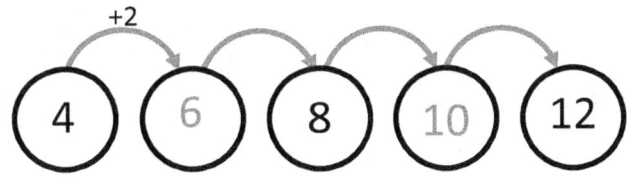

9. Count by 2 from 1 to 11

10. Count by 2 from 9 to 19

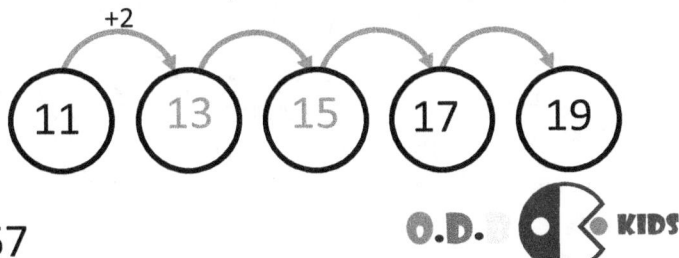

Identifying even / odd numbers 1-20

Grade 1 Counting & Numbers Worksheet

Circle the EVEN number(s).

1) (10) 9 (8) 5 (2) (4)

2) 9 (10) 7 (18) (20)

3) 3 (12) 11 (20) 7 1

4) (2) 5 17 13 11

5) (20) 17 5 (6) (2) 3

6) (10) (8) 3 (2) 1

7) 16 14 11 2 1 7

8) (14) 7 5 (18) 19

9) (10) 9 (14) 15 3

10) (2) 13 (16) (12) (6)

Circle the ODD number(s).

11) (9) 10 (7) 18 20

12) (3) 10 2 8 (9)

13) 2 (5) (17) (13) (11)

14) (15) 8 10 (17) (7)

15) 10 8 (3) 2 (1)

16) 14 16 (9) 12 (5)

17) 14 (7) (5) 18 (19)

18) 20 10 (11) 8 (9)

19) 2 (13) 16 (11) 6

20) (13) (17) 6 2 20

Identifying even / odd numbers 1-20

Grade 1 Counting & Numbers Worksheet

Circle the EVEN number(s).

1) (10) 17 (8) 5 3 (4)
2) 9 1 (8) 13 (20)
3) 3 12 11 20 7 8
4) (2) 5 (16) 13 (12)
5) 7 16 5 6 2 3
6) 7 (8) 5 (2) 1
7) (18) 15 11 3 (16) 7
8) (14) 7 19 (18) (20)
9) 13 9 (14) 15 (4)
10) (2) 7 1 (12) (6) (4)

Circle the ODD number(s).

11) 4 (5) (7) (1) 20
12) (3) 20 2 (5) (9)
13) 2 (5) 8 (13) (15)
14) (15) 12 (5) (17) 6
15) 6 8 (3) 2 (17)
16) 14 (17) 2 12 (5)
17) 14 (7) 8 18 (19)
18) (3) 10 (11) 8 (9)
19) 2 (13) 16 (19) 6
20) (9) (17) 6 2 20

Identifying even / odd numbers 1-20

Grade 1 Counting & Numbers Worksheet

Circle the EVEN number(s).

1) 17 19 15 1
2) 9 (8) 15 1
3) (12) 5 3 (10)
4) (18) 15 11 5
5) (12) (6) 1 19
6) (20) (10) 7 5
7) 11 17 (14) (8)
8) 1 19 (2) (20)
9) 3 (6) 15 5
10) 15 5 (20) (8)

Circle the ODD number(s).

11) (1) 19 16 6
12) (19) (9) 2 (3)
13) 16 14 (13) 12
14) 6 (11) (17) (13)
15) 18 4 (19) (9)
16) 14 8 20 6
17) 8 4 12 18
18) (17) 2 (15) (11)
19) (7) 10 (15) 2
20) (7) (9) 20 2

Identifying even / odd numbers 1-100

Grade 1 Counting & Numbers Worksheet

Circle the EVEN number(s).

1) 5 1 (82) (50)
2) (48) 47 (4) (86)
3) (60) (6) (52) 53
4) (6) 4 (2) 7
5) 19 (6) 7 (66)
6) 85 45 97 (6)
7) (50) (4) 2 (74)
8) (2) (8) (60) (62)
9) (6) (2) 3 (62)
10) 41 (52) 7 (92)

Circle the ODD number(s).

11) (91) 78 26 (9)
12) (35) (1) (7) 64
13) (53) (39) 38 94
14) 32 80 (91) (57)
15) 2 (31) (97) 68
16) (55) (7) (17) 8
17) 96 92 68 (1)
18) 4 6 38 8
19) 58 96 (27) (5)
20) (17) 100 (1) (3)

71

Identifying even / odd numbers 1-1000

Grade 1 Counting & Numbers Worksheet

Circle the EVEN number(s).

1) (64) 587 (478) (52)
2) 61 35 303 247
3) 3 (988) 261 1
4) 665 930 318 268
5) 9 (104) (2) (914)
6) 437 2 41 348
7) (960) (474) 5 (20)
8) 960 266 456 55
9) (2) (8) (62) (898)
10) 898 1 9 7

Circle the ODD number(s).

11) 6 (687) (181) 880
12) (299) 36 (53) 988
13) (7) (815) 416 (807)
14) 350 36 (687) 152
15) 98 22 (981) (5)
16) 50 (7) (523) (659)
17) (23) (397) 188 (567)
18) (759) 76 32 (837)
19) 20 3 5 17
20) (43) 548 (29) 2

Numbers as words (0-20)

Grade 1 Numbers Worksheet

Circle the correct number for each word.

eight	5	13	**(8)**
sixteen	**(16)**	6	19
fourteen	**(14)**	24	4
twenty	2	12	**(20)**
ten	9	**(10)**	2
three	**(3)**	6	9
thirteen	16	**(13)**	4
nineteen	9	16	**(19)**
eleven	**(11)**	12	1
twelve	**(12)**	11	1

Numbers as words (1-20)

Grade 1 Numbers Worksheet

Draw a line between the number to its word.

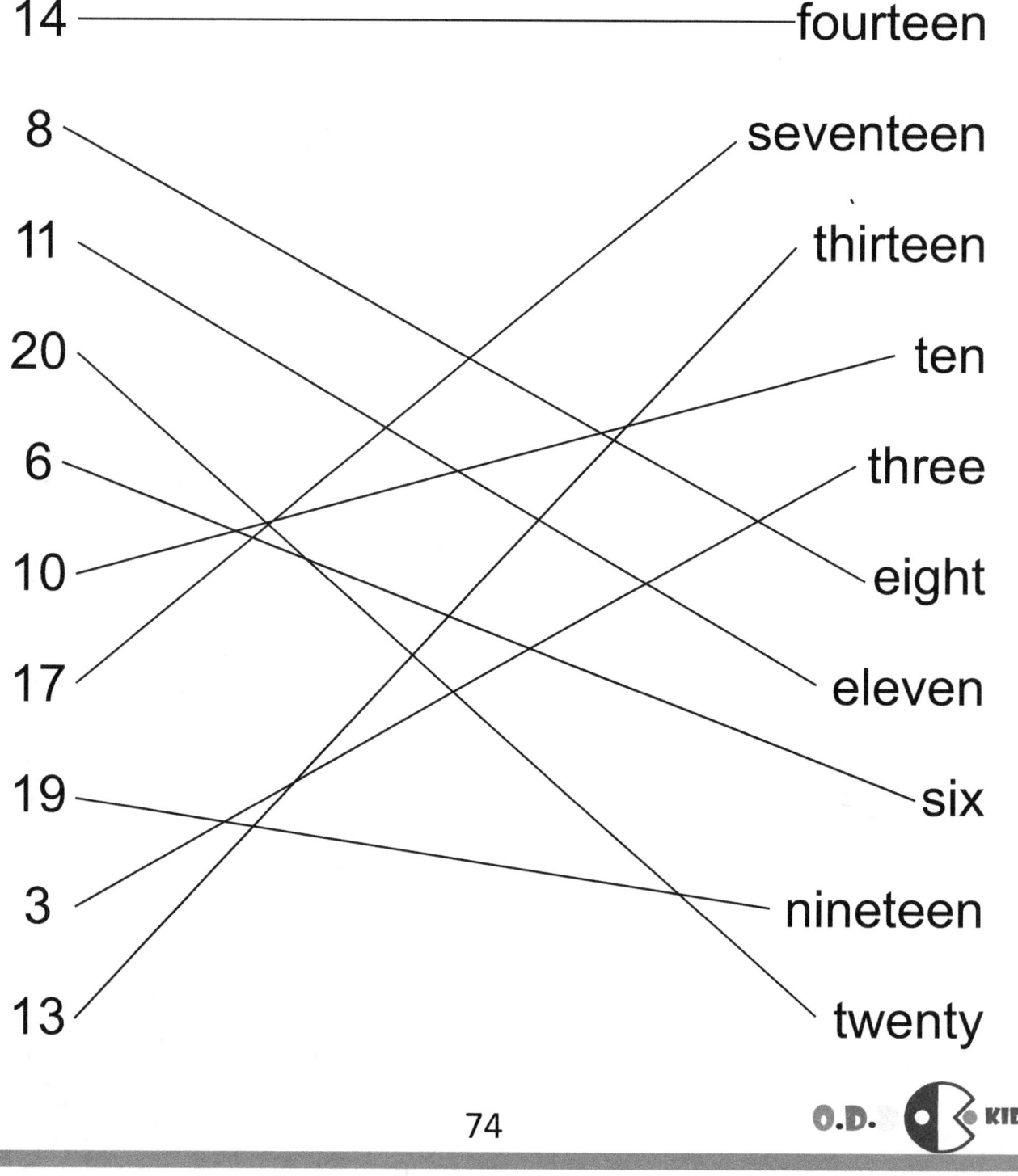

Numbers as words (0-20)
Grade 1 Numbers Worksheet

Circle the correct number for each word.

twenty-four	14	(24)	4
twenty	(20)	2	16
thirty	13	21	(30)
twelve	10	11	(12)
sixteen	(16)	13	19
twenty-nine	23	26	(29)
eight	16	(8)	4
fourteen	24	16	(14)
twenty-one	22	12	(21)
twenty-seven	7	20	(27)

Numbers as words (1-20)

Grade 1 Numbers Worksheet

Draw a line between the number to its word.

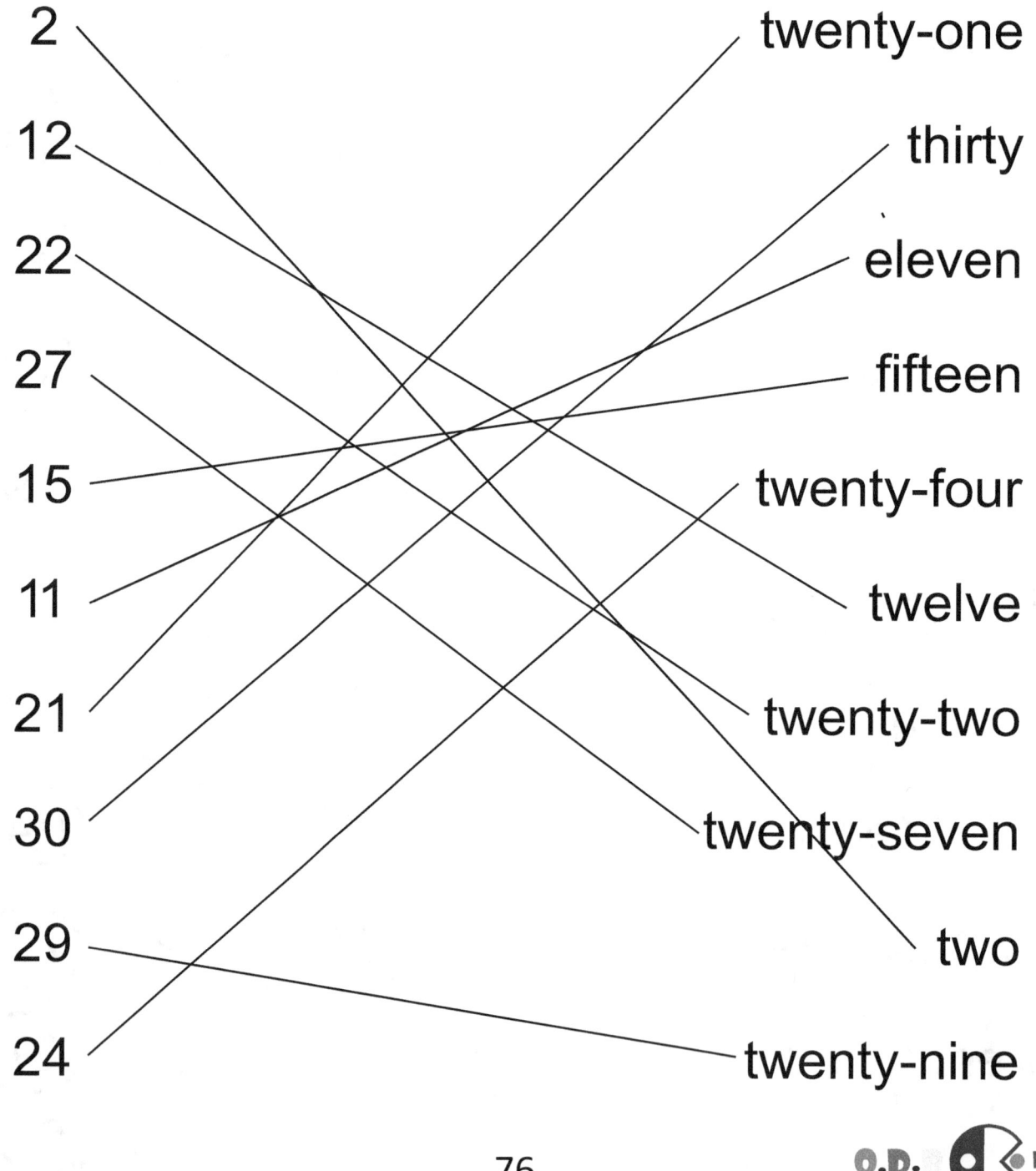

Numbers as words (0-20)
Grade 1 Numbers Worksheet

Circle the correct number for each word.

eighty-four	44	88	**(84)**
fifty-six	**(56)**	65	66
thirty-three	13	23	**(33)**
one hundred	101	**(100)**	10
fifteen	**(15)**	55	25
ninety-nine	66	96	**(99)**
eighty	68	**(80)**	40
one hundred twelve	**(112)**	61	114
twenty-one	29	12	**(21)**
sixty-seven	61	68	**(67)**

Numbers as words (1-20)

Grade 1 Numbers Worksheet

Draw a line between the number to its word.

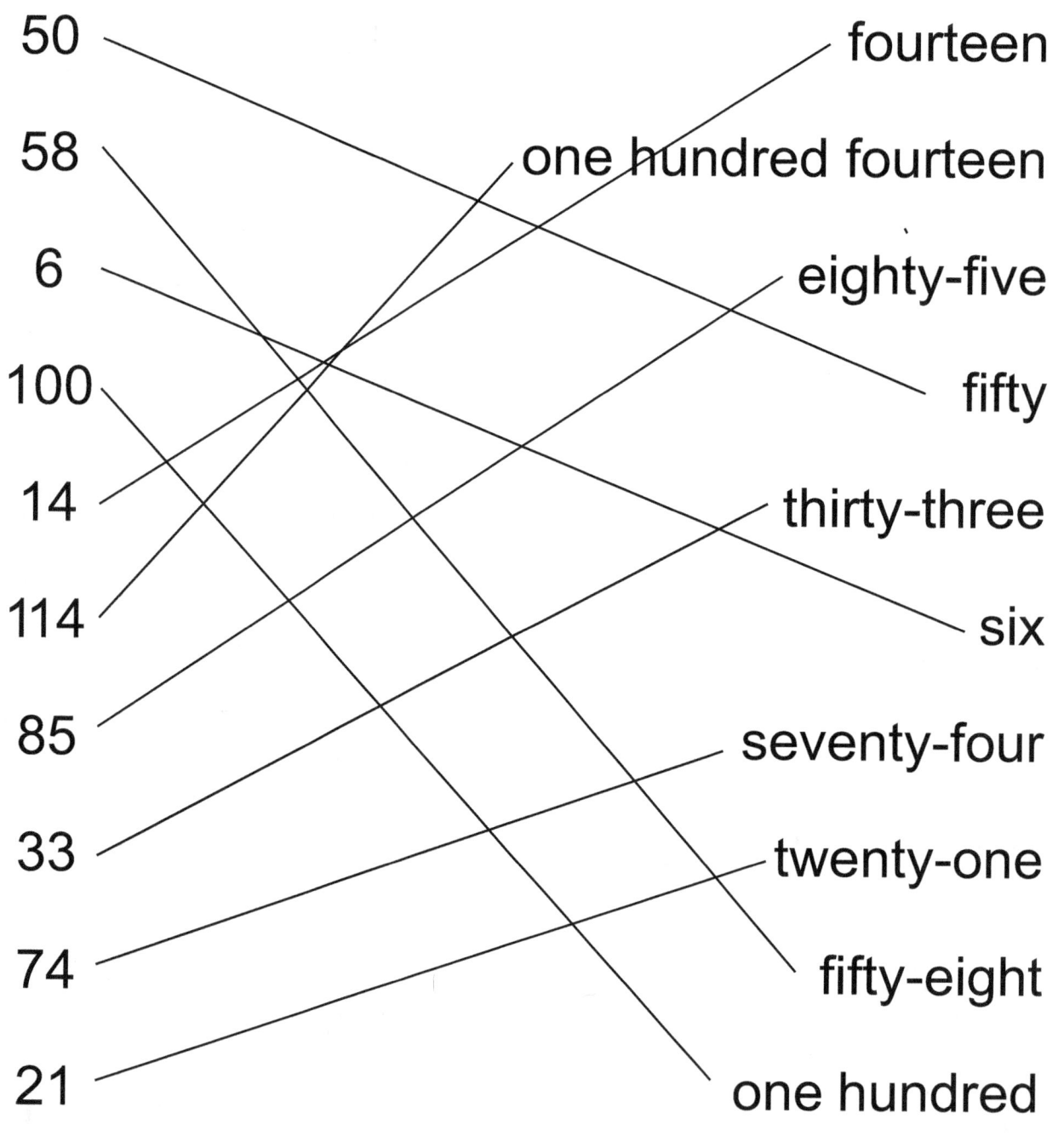

Identifying tens and ones

Place Value

Fill in the correct tens and ones for the given numbers.

tens 8 and ones 6 = 86

tens 1 and ones 6 = 16

tens 3 and ones 6 = 36

tens 2 and ones 5 = 25

tens 7 and ones 6 = 76

tens 1 and ones 4 = 14

tens 6 and ones 3 = 63

tens 1 and ones 7 = 17

tens 2 and ones 3 = 23

Identifying tens and ones

Place Value

Fill in the correct tens and ones for the given numbers.

tens	3	and	ones	0	= 30
tens	2	and	ones	5	= 25
tens	4	and	ones	6	= 46
tens	7	and	ones	0	= 70
tens	8	and	ones	9	= 89
tens	7	and	ones	3	= 73
tens	1	and	ones	9	= 19
tens	3	and	ones	7	= 37
tens	9	and	ones	4	= 94

Identifying tens and ones

Place Value

Fill in the correct tens and ones for the given numbers.

tens	4	and	ones	2	= 42
tens	6	and	ones	7	= 67
tens	1	and	ones	3	= 13
tens	9	and	ones	3	= 93
tens	6	and	ones	4	= 64
tens	5	and	ones	7	= 57
tens	7	and	ones	2	= 72
tens	1	and	ones	6	= 16
tens	9	and	ones	2	= 92

Identifying tens and ones

Place Value

Fill in the correct tens and ones for the given numbers.

tens	7	and	ones	8	= 78
tens	2	and	ones	9	= 29
tens	3	and	ones	7	= 37
tens	6	and	ones	3	= 63
tens	7	and	ones	6	= 76
tens	9	and	ones	4	= 94
tens	1	and	ones	7	= 17
tens	3	and	ones	8	= 38
tens	1	and	ones	8	= 18

Identifying tens and ones

Place Value

Fill in the correct tens and ones for the given numbers.

tens 9	and	ones 1	= 91
tens 2	and	ones 6	= 26
tens 3	and	ones 7	= 37
tens 1	and	ones 2	= 12
tens 8	and	ones 8	= 88
tens 9	and	ones 7	= 97
tens 3	and	ones 4	= 34
tens 5	and	ones 0	= 50
tens 5	and	ones 7	= 57

Identifying tens and ones

Place Value

Fill in the correct tens and ones for the given numbers.

tens	1	and	ones	5	= 15
tens	6	and	ones	7	= 67
tens	9	and	ones	4	= 94
tens	3	and	ones	6	= 36
tens	4	and	ones	3	= 43
tens	4	and	ones	9	= 49
tens	8	and	ones	4	= 84
tens	2	and	ones	2	= 22
tens	3	and	ones	3	= 33

Combining tens and ones

Place Value Worksheet

Fill in the correct tens and ones for the given numbers.

24 = 2 tens and 4 ones

12 = 1 ten and 2 ones

67 = 6 tens and 7 ones

96 = 9 tens and 6 ones

55 = 5 tens and 5 ones

80 = 8 tens and 0 ones

39 = 3 tens and 9 ones

48 = 4 tens and 8 ones

■ an example of

2 tens and 2 ones

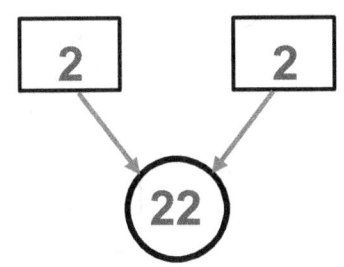

Combining tens and ones

Place Value Worksheet

Fill in the correct tens and ones for the given numbers.

| 54 | = | **5 tens and 4 ones** |

| 23 | = | **2 tens and 3 ones** |

| 67 | = | **6 tens and 7 ones** |

| 81 | = | **8 tens and 1 one** |

| 36 | = | **3 tens and 6 ones** |

| 75 | = | **7 tens and 5 ones** |

| 42 | = | **4 tens and 2 ones** |

| 10 | = | **1 ten and 0 ones** |

Combining tens and ones

Place Value Worksheet

Fill in the correct tens and ones for the given numbers.

| 83 | = | **8 tens and 3 ones** |

| 78 | = | **7 tens and 8 ones** |

| 14 | = | **1 ten and 4 ones** |

| 33 | = | **3 tens and 3 ones** |

| 41 | = | **4 tens and 1 one** |

| 60 | = | **6 tens and 0 ones** |

| 96 | = | **9 tens and 6 ones** |

| 25 | = | **2 tens and 5 ones** |

Combining tens and ones

Place Value Worksheet

Fill in the correct tens and ones for the given numbers.

20	=	**2 tens and 0 ones**
79	=	**7 tens and 9 ones**
51	=	**5 tens and 1 one**
47	=	**4 tens and 7 ones**
63	=	**6 tens and 3 ones**
38	=	**3 tens and 8 ones**
86	=	**8 tens and 6 ones**
14	=	**1 ten and 4 ones**

Combining tens and ones

Place Value Worksheet

Fill in the correct tens and ones for the given numbers.

| 37 | = | 3 tens and 7 ones |

| 28 | = | 2 ten and 8 ones |

| 40 | = | 4 tens and 0 ones |

| 91 | = | 9 tens and 1 one |

| 75 | = | 7 tens and 5 ones |

| 64 | = | 6 tens and 4 ones |

| 54 | = | 5 tens and 4 ones |

| 87 | = | 8 tens and 7 ones |

Combining tens and ones

Place Value Worksheet

Fill in the correct tens and ones for the given numbers.

| 58 | = | 5 tens and 8 ones |

| 66 | = | 6 tens and 6 ones |

| 19 | = | 1 ten and 9 ones |

| 37 | = | 3 tens and 7 ones |

| 43 | = | 4 tens and 3 ones |

| 87 | = | 8 tens and 7 ones |

| 94 | = | 9 tens and 4 ones |

| 26 | = | 2 tens and 6 ones |

MY KINDERGARTEN MATH SKILLS ACTIVITY WORKBOOK

- Copyright 2020/2021 ©-

Copyright of this website and its contentskharaz for Printing . All rights reserved.

Redistribution or reproduction of part or all of the content in any form is prohibited

www.ingramcontent.com/pod-product-compliance
Lightning Source LLC
Chambersburg PA
CBHW062359220526
45472CB00008B/1873